# ROOFS

# E~~SSENCE~~ ~~BO~~OKS ON BUILDING

*General Editor:* J. H. Cheetham, ARIBA

Other titles in the Essence Books on Building Series

# ROOFS

**Roy E. Owen** Dip. Arch., ARIBA

*Senior Lecturer, Department of
Architecture, Oxford Polytechnic*

# MACMILLAN

SBN 333 12993 8

695

*Published by*

THE MACMILLAN PRESS LIMITED

*London and Basingstoke*
*Associated companies in New York*
*Melbourne Toronto Dublin*
*Johannesburg and Madras*

To Esme

*Printed in Great Britain by*
*The Whitefriars Press Ltd., London and Tonbridge*

# Preface

This book in the Essence series deals with the functional basis, structural principles and assembly and erection of roof systems, from the traditional timber frameworks still in common use, to the new forms used to cover large uninterrupted spaces.

Recognition is given here to the systems and components which, within the last 10 to 15 years, have been developed and produced by manufacturers specializing in a limited range of structural techniques, usually employing one principal material. Examples of this trend are light steel space decks based on small standardized pyramidal units, and timber nail-plate trussed rafters, made to a range of spans and pitches, for common coverings and loading. To a large extent the process of selecting a roof form has been revolutionized, as the designer now has a vocabulary of available components that may be compared on a basis of functional, visual and economic suitability. Commercial plywood web and light lattice steel joists have extended spans previously limited by the capability of sawn timber sections.

Most of the roofs described here are characteristic of current practice. Those which have become obsolete or have been largely superseded are omitted. Certain forms referred to are still under development and may be in wider use in the future. By definition, this book gives the essence of the subject, and is not intended primarily for readers requiring information for the detailed design of roof structures. References are given as a guide to further investigation and study. It will be self-evident that decisions on the form, nature and dimensions of large-span and complex roofs should only be taken by the architect working in continuous consultation with the structural engineer, and the fabricator, constructor and costs adviser wherever appropriate. It is in the nature of modern construction that the early collaboration of many specialists is essential if such a major building element is to be efficient, in terms of structure, erection technique, cost and functional performance.

*In situ* and precast reinforced concrete slabs of short and medium span are often employed in the same building for both floors and flat roofs, in flats and office blocks, for example. In such cases the roof differs only in its design for imposed loading, and in the application of thermal insulation and waterproofing. The Essence book *Floors* deals with these slabs, and therefore reference to them in the present volume is confined to lightweight precast reinforced concrete units produced principally for roof decking. For the same reason roof coverings will be referred to only where there is a direct relationship between the covering and the form or function of elements of the main roof structure.

Conventionally, the roof may be regarded as the building element acting as a cover between bounding walls or columns, having clearly limited surfaces and perimeter. This definition is confused by the fact that some buildings have walls and roof in continuous transition. Examples are ancient beehive huts whose walls corbel inwards, terminating in an apex, and in modern practice, shell domes and curved arch structures which rise from ground level. These continuous enclosures are included here, together with frameworks whose supporting columns are an integral part of the roof structure.

*Dimensions*
Dimensions are given in metric SI units. Where a metrically sized member has been produced at the time of publication, this will be given. Other component sizes will be those currently available, expressed in metric terms.

*Building Regulations*
Where statutory regulations affecting the design and selection of roof components are referred to in the text, these are The Building Regulations 1965 and amendments, applicable in England and Wales. Equivalent regulations affecting Scotland will be found in The Building Standards (Scotland) Regulations 1970. The London Building Acts — Constructional By-laws (1965) cover the Inner London Boroughs and the City.

*Acknowledgements*
The author wishes to acknowledge the many authorities, manufacturers and associations who have contributed advice, information and photographs, and to thank particularly Hal Cheetham, ARIBA, general editor of this series, for his detailed advice on the preparation of the text and illustrations. The line drawings are by the author.

# Acknowledgement
# for Illustrations

# Contents

Grid dome, Biosphere. Expo 67, Montreal

# Introduction

This book is arranged in three sections. Section One deals with the *Functions and Performance Requirements* of roofs, and with factors governing the selection of roof structure. Sections Two and Three divide roof forms into two categories, *Skeletal Structures* and *Surface and Membrane Structures* respectively. Most roofs can be allocated to one of these divisions; a few embody both conceptions. For example, a roof deck may combine pyramidal stressed-skin units with linear metal ties (Fig. 21k). On the other hand northlight shells may have cylindrical or conoid surfaces, with perforated deep beams or lattice girders as edge supports (Fig. 27e and f). Figures 1 and 2 illustrate the differences between skeletal and surface structures in roofs of similar cylindrical vaulted geometry.

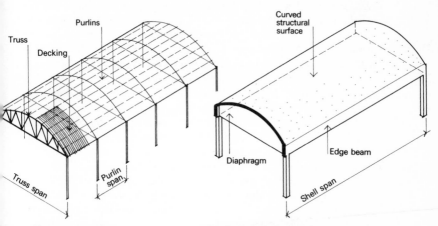

Figure 1. Skeleton structure.
Bowstring truss

Figure 2. Surface structure.
Single bay long span cylindrical shell

In a skeleton or linear structure, the length of each member is great in relation to its width and depth. The members are assembled into a network, receiving the loads imposed by the covering, which include imposed wind and snow loads, and transmitting them to supporting walls, columns or foundations. These members were originally found in naturally-occurring timber, but are now those most convenient for the conversion, manufacture, transportation,

1

handling and assembly of the material employed. Timber may be sawn or planed and steel formed in hot- or cold-rolled sections. We also have extruded aluminium-alloy profiles, and precast reinforced concrete members. In range of scale and size, there are battens, rafters, joists, struts, ties and beams. Assembly and jointing techniques vary from the simple and relatively inefficient nailing of adjacent timbers, to the use of complex multi-way connectors for grid networks, through which the stresses in incoming members are transmitted. In the simple domestic timber roof, a tradition of member-sizing and assembly normally ensures an adequate and stable structure. The long-span grid network requires advanced and specialized structural design method, often with model testing.

Within the skeletal category further sub-divisions occur, into single-directional forms, in which a joist, beam or truss transmits loads in one direction to supports (Fig. 13), and multi-directional grids and decks, distributing load in two or more directions, on to a perimeter wall or beam, or on to widely-spaced columns (Fig. 21c). In recent years there have been considerable developments in the design and fabrication of long-span decks and grids, in both purpose-designed and standardized forms.

Skeletal structures do not provide cover. This is given by a combination of materials, for waterproofing and insulation, and for a minor spanning function on to support members at suitable centres. The loads on surface and membrane structures are distributed in many directions, to points of collection and support, within the structural roof cover itself (Fig. 2). A fabric tent is a simple tensioned membrane; a long-span curved concrete shell is a surface structure, mainly in compression. In both cases the stress conditions are favourable to the structural and enclosing material.

A plywood sheet roof decking and a two-way spanning *in situ* reinforced concrete roof slab both qualify as surface structures, but as spans increase above the optimum for each material, increased thickness, deadweight and material usage make them inefficient and limit their capability. An increase of structural depth of the whole unit by curvature, folding or corrugation is then necessary in order to create a beam effect and by this means a basically thin material may be used over spans which are large in relation to the thickness of the structural surface. A corrugated metal deck, spanning about 3 m between purlins, is a minor example of this technique (Fig. 12k). A major one is the case of a cylindrical reinforced concrete shell having a clear span of 30 m with a crown thickness of less than 100 mm.

Although surface structures possess the advantage of an inherent covering function, the thin membrane of a dense material has inadequate thermal insulation for most building functions and climatic conditions. An insulant, applied to the upper structural surface to reduce movement due to climatic temperature variation, requires waterproof covering. The soffit of the roof may also require treatment with a low-density material in order to reduce sound reflection, which is particularly troublesome where an acoustic focus occurs under a concave concrete surface. It will be seen that the advantage possessed by surface structures of providing cover is only one aspect of the performance of the roof in controlling the internal environment.

# SECTION ONE
# Functions and performance requirements

This Section reviews the inherent or incorporated characteristics in roofs, which give them appropriate performance in contributing — with the other external building elements — to the provision of adequate durability and a controlled internal environment.

These characteristics, dealt with in Chapters 1 and 2 respectively, are:

| | |
|---|---|
| Durability | Sound insulation |
| Weather resistance | Fire protection |
| Strength and stability | Adaptability and modification |
| Thermal insulation | Maintenance, demolition and renewal |
| Thermal movement | Economic factors |

# Chapter 1

## Durability

Efficient in excluding water while it remains unperforated, and of some effectiveness in providing privacy, screening from direct sun, shelter from wind and reduction of thermal loss, a roof may be simply constructed of stretched woven proofed fabric, as in holiday and circus tents. These "buildings", of relatively low capital cost, are suitable for intermittent use over fairly long periods, but are of limited durability in permanent exposure to severe weather. Such single-skin structures offer poor sound reduction, low thermal insulation and no resistance to fire. They may remain intact in high winds, if correctly erected, and will suffer quite large distortions without visible detriment or impaired stability.

That ruins are almost invariably roofless is an indication of the severity of exposure of the roof covering, and the effects of stresses and decay over long periods. Outstanding examples from antiquity, such as the Pantheon in Rome, the Gallarus Oratory on the Dingle Peninsula in western Ireland, and St Kevin's Church, Glendalough, co. Wicklow, have survived for more than a thousand years, being roofed with masonry.

Failure of waterproofing will initiate deterioration and ultimately lead to loss of structural integrity. Such failure may be caused by:

1. *Decay of a non-durable material* such as thatch, or the lamination and breakage of small roofing units, e.g. tiles, slates and shingles, caused by exposure to sun, temperature change and frost.

2. *The disturbance or stripping of coverings by wind pressure or suction*, in severe exposure, or where fixings are inadequate or have corroded.

3. *The splitting of sheet and continuous-membrane coverings, by movement of the substructure*. This movement should be restricted by limiting structural deflection and thermal movement, and by isolation of the covering from its substrate, allowing independent movement of covering and support.

4. *Splitting or holing of a continuous membrane following bubbling due to water vapour pressure*, produced by solar heating of water contained within the roof structure. The water may be residual water used in casting concrete or in mixing wet screeds, rain falling

7

upon and being absorbed by the structure prior to the application of the waterproofing, or condensation (within the roof structure) of water vapour migrating from the air beneath it. Blistering leads to water penetration only when splitting occurs, but it is unsightly on any visible roof.

*Reduction of water content in roofs*

The incidence of water within the roof structure may be reduced by:

(a) Adopting dry forms of construction, such as matured precast reinforced concrete roof slab units (Fig. 5c), metal decking or other roof decks in which water needed in manufacture has largely been dried out before delivery.

(b) Using insulants and methods of forming falls that do not require water in mixing or application.

(c) Draining residual construction water, or water accumulated from rainfall on an exposed slab, through weepholes formed at low points.

(d) Covering the roof with a waterproof membrane as soon as possible after construction. This may require careful phasing of the waterproofing specialist's work by the general contractor, the employment of a single specialist for decking, insulation and covering and responsible for co-ordinating the three operations, or by the use of decking or insulation slabs covered by the waterproofing or its primary layer in the factory.

(e) The continuous ventilation of entrapped moisture through outlets in the roof covering. Systems have been produced commercially for venting the underside of the covering through freestanding ventilators, or behind upstand skirtings and under flashings, where wall to roof abutments or parapets occur. Efficiency depends on the freedom and speed with which water may travel across the roof area to and through such outlets.

*Deterioration of materials in roofs*

Condensation occurring within the roof structure will rot organic insulation materials, such as fibreboard, strawboard and cork slab, and may lead to water dripping from the roof soffit or the ceiling below. Preventive treatment consists of providing a high level of thermal insulation, and an efficient, continuous vapour barrier at or near the underside of the roof system. Buildings in which large volumes of steam are discharged into the interior by an industrial process may require mechanical ventilation, heating of the roof void above ceiling level, efficient thermal insulation and the provision of a vapour barrier if condensation on ceiling surfaces and within the roof structure is to be prevented.

The problem of condensation in roofs is discussed in detail in

Building Research Station Digest no. 51 (Second Series), 1964, *Developments in Roofing.*

Water, derived from natural precipitation, from construction processes, or from condensation, causes deterioration in roof fabric by establishing rot or corrosion. The rotting of organic insulation reduces its thermal insulation value and may lead to staining of ceilings. Water will not necessarily initiate rapid breakdown in a supporting structure of reinforced concrete or adequately protected metal, but deckings of non-durable hardwoods and softwoods, compressed strawboard or woodwool will decay and soften, and structural integrity may ultimately be threatened. An increase in the moisture content of non-durable structural roof timbers, caused by roof leakage or lack of ventilation, may lead to dry rot *(merulius lachrymans)* with subsequent roof failure.

Water penetrating a defective roof covering may be conducted to a point remote from its entry, and will not be detected unless the roof is open and readily inspected, or until dampness becomes visible.

Damp conditions conducive to decay also occur in timber roofs with sound covering which are warm and unventilated. Flat and mono-pitched joisted roofs, with porous ceilings fixed directly to the underside of the joists, are vulnerable. A continuous vapour barrier at ceiling level, adequate perimeter ventilation of the roof space, and thermal insulation (preferably at deck and ceiling positions) will help to prevent trouble, although preservative treatment of all roof timber is always a desirable safeguard (Fig. 3).

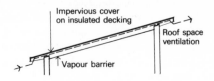

Figure 3. Avoidance of condensation within roof space

Pitched timber roofs with slating or tiling over sarking felt normally have adequate natural ventilation through gaps in the felting and at eaves. However, if the covering is close and continuous (metal sheet or bituminous felt, for example), migration of warm moist air towards the roof apex will create damp conditions there, conducive to rot. This has often occurred in the roof timbers of old lead-roofed churches after installation of central heating, when water in the masonry has dried out into the interior, and has risen into the unventilated roof. Occupiers of houses built by traditional "wet" methods should (particularly if first occupation occurs in autumn or winter) initiate central heating gradually, with ventilation of rooms, to reduce the amount of construction water that may reach the roof space.

Most softwoods used in roof structure are vulnerable to insect as well as to fungal attack, and should be treated with a preservative applied by pressure impregnation or by effective immersion. The increase on the basic cost of softwood for this treatment is approximately 12½%. In some areas the incidence of insect attack is very severe, and the Building Regulations 1965 require preservative treatment of softwoods within roofs in parts of Surrey and Middlesex to prevent infestation by house longhorn beetle. Regulation B4 states the requirement, B5 the treatments, and Schedule 4 (with Amendment 12 of The Second Amendment 1966) gives the local authority areas to which B4 applies.

The penetration of water into dense reinforced concrete roofs is not likely to cause deterioration, as the dense material excludes the oxygen and carbon dioxide necessary for reinforcement corrosion. Lightweight concretes based upon the use of low-density aggregates, or formed by the creation of gas bubbles within the mix, allow entry of damp air, favouring corrosion. The reinforcement in lightweight or very thin concrete should be protected by zinc or bitumen coating. (See B.R.S. Digest no. 109 (Second Series), 1969.)

In terms of economy and structural purity it is attractive to consider the exposure of dense reinforced-concrete roofs, without additional waterproofing. Although the use of concrete in compression, with or without prestressing, tends to reduce the occurrence of cracks, any shrinkage or stress cracks, or "honeycombing" voids due to imperfect consolidation of the wet mix allow leakage, soffit staining and steel corrosion. Moreover, exposure of the structural surface without intervening insulation or solar reflectors will result in large thermal movements, related to roof area and climatic temperature range. In dry, equable climates, an unprotected concrete shell canopy may be used over such structures as grandstands, where insulation and total water exclusion are not of paramount importance. For most buildings, however, insulation and waterproof finishes are essential.

Although inherent durability of roof structure is a useful attribute, non-durable or unprotected materials will be satisfactory if the covering is maintained in watertight condition, by inspection, repair or restoration, as may be necessary. For the roofs of buildings which house industrial processes producing steam and corrosive chemicals, dense reinforced concrete or naturally durable timbers are suitable basic materials, depending, of course, on the nature and degree of the exposure. Deckings used in conjunction with skeletal structures must be assessed for their ability to resist ambient conditions.

Structural steelwork will be subject to corrosion unless the members and fixings are effectively painted, or are protected by zinc, applied by galvanizing (immersion of the component in molten zinc) or zinc spraying (in which the metal, vaporized in a gas flame, is projected on to the treated part). A satisfactory paint system on steel requires a clean and dry metal surface, the application of a rust-inhibitive primer and an oil or bitumen-based paint. (See Building Research Station Digest no. 70 (Second series).) Corrosion will begin at points where the paint has been omitted, or damaged, or has flaked off. As repainting of the whole roof assembly may be required at intervals, complex steel roof networks are undesirable for aggressive industrial atmospheres, unless a durable unbroken coating can be guaranteed to provide long-term durability without maintenance. Access for repainting to multi-directional skeletons (particularly if angles, tees and I sections are used), is difficult, and application laborious. Simple portal frame and purlin roofs, using large I sections, or sealed tubes, are more satisfactory.

*Weather resistance*
The weather resistance of a roof as a whole is principally a function of its external covering, but the covering may depend for its effectiveness on the ability of the structure to provide the required pitch or falls. In the case of thatch, for example, a steep pitch is essential to enable water to travel downward and towards the eaves without penetrating its thickness.

Small tiles and slates, with multiple vertical butt joints, also demand a steep pitch for water shedding. In conditions of wind pressure, penetrating water or fine snow is received by the reinforced bituminous sarking felt which underdraws the covering and is draped over the rafters beneath the battens. The sarking conducts this water into eaves guttering.

Large patent interlocking concrete slates may be laid at pitches as low as 15 deg, and have overlap joints designed to prevent water penetration. At lower pitches, large corrugated sheets span 1½ to 3 m over bearings, by virtue of the structural depth imparted by their corrugations. Sealing depends upon the edge corrugations overlapping, and upon adequate end laps between sheets. At low pitches mastic cord is used to seal the end laps, and at *very* low pitches special corrugated sheets have end laps shaped to increase pitch at the overlap. Asbestos cement, aluminium alloy and protected steel sheets are commonly used. Aluminium is available in very long lengths to eliminate the need for end laps on most roofs.

The roof structure must provide appropriate pitch for the run-off of water. The structural design of a triangular lattice truss is generated by the inclination of the top chord. Favourable profiles for water dispersal are provided by many other structural roof forms. Roof slabs, which are necessarily laid dead flat, require the application of a layer that will provide adequate falls to outlets. It will be apparent that support structure, deck and covering are closely inter-related.

For "flat" roofs, beams of constant section require tapered *firring pieces* (Fig. 19a) to provide the fall, or the beam itself may be inclined. The design of nominally parallel trusses may be modified to incorporate a fall, or the whole truss may be cambered, cranked or tapered to provide the fall requirements of a sheet or jointless covering. Again, the whole truss may be pitched. Lattice roof grids and decks (Fig. 21c, e) can be cambered during erection, by adjustment of the lower ties at the nodal connections.

A complication is the need to provide outlets for rainwater within the roof area or around its perimeter. In the stressed-cable roof (Fig. 26) and the hyperbolic-paraboloid roof (Fig. 34) water is conducted to low points for collection into large hoppers or gratings. On flat roofs the outlets should ideally be placed on a square or near square grid, so that falls are identical, and the firring or screeding is kept to a minimum, commensurate with efficient water clearance. This is often made difficult to achieve in practice by irregular plan outline and by plan variations on intermediate floors between outlet and drain connection.

The use of cement-bound screeds at considerable thicknesses should be avoided. A screed to provide a fall of 1 in 80 over only 12 m would have to be at least 150 mm thick. Thick screeds are undesirable because:

(a) The cost of materials and labour increases with thickness; with lightweight insulating screeds this is more marked than with cement-and-sand screed.

(b) The deadweight of the maximum thickness may increase structural sizes and reinforcement.

(c) A large volume of water for mixing will be required. Much of this will be trapped when the waterproof covering is laid. Drying out will be slow, even when a special ventilation system is incorporated. (See BRS Digest no. 51 (Second Series).)

Normal Portland cement-and-sand screeds should not be laid less than 20 mm thick. Lightweight screeds used for thermal insulation must not be of less thickness than that required for an adequate insulating function — a minimum of 40 mm. Maximum and minimum thicknesses of all screeds should be shown on drawings. Arrows showing the general direction of roof falls are not sufficient.

Wet screeds should be used only on "heavy" roof slabs, and not on lightweight decks.

The use of expanded-plastics slabs, taper-cut to provide falls, overcomes the problems of deadweight and water retention, but cost increases with thickness. With these very efficient thermal insulators, more than 40 to 50 mm is not normally required. (See BRS Digests nos. 93 and 94 (Second Series).)

Where the beams or trusses are exposed internally the provision of falls in "flat" roofs by means of firring, canting or cambering is often visually objectionable. Where windows extend to the underside of beams, a level junction is usually visually and technically desirable. These problems, with the need for economy, and to avoid screeding, have led to the provision of roofs which are, theoretically, perfectly flat. The assumption is that on a level surface, water will flow through any outlets flush with or below this surface, and that these outlets need not be regularly spaced. Unfortunately, inaccuracies in workmanship, deflection of the roof (which may take place over a long period), and ridges formed at sheet-roof junctions prevent the achievement of true flatness in practical terms. Stone chippings, used as a solar reflector or for appearance, and fine debris also inhibit water flow. Minimum recommended falls usually allow for some loss of effective fall, but should be increased whenever possible.

Random pools of water on the roof surface will be unsightly, seen from above. Strong evaporation by solar heating cools the wetted area and stress in the roof covering may occur at the junction with the heated dry area. Standing water rots the vegetable (jute) fibre in fibre-based roofing felt, although the cooling tends to preserve the bitumen. If a leak occurs under a pool, water clearance is difficult. There may be scouring by ice and detachment of bonded chippings when pools freeze in winter.

The specifically-designed flooding of flat roofs may be adopted to assist in achieving a stable internal temperature in the building, but requirements are exacting; an adequate artificial water supply, an overflow to control the maximum depth of water, falls to valved outlets to permit complete clearance, a roof designed for the dead load, and a perfect waterproof covering with deep perimeter skirtings must all be taken into account. Thermal insulation under the roof covering is required to prevent condensation on the supporting slab, owing to the cooling effect of the water. The water would normally be drained off in winter, to avoid lateral thrust from ice on the containing parapet walls or beams.

## Strength and stability

The structural elements and components of a roof must be designed with adequate strength for the roof to remain intact, stable and with limited controlled deformation under working stresses, due to dead and imposed loads.

*Dead loads* are the self-weight of the roof structure and its claddings and any other static loads from equipment carried on or suspended from the structure. Over short and medium spans structural sizes are small and dead weight is usually not critical, but on longer spans the structure may become very heavy in relation to cladding and imposed loads. It is therefore important to reduce this ratio wherever possible, by the use of light claddings and finishings, refined structural techniques, the use of light structural materials such as aluminium alloy (where cost permits), or design forms that produce thin structural surfaces or membranes.

Structural design procedures vary with span and roof form. The sizes of timber members for a simple roof may be derived from traditional examples or from tables in the Building Regulations. Structural data sheets for timber roof trusses of short and medium spans are published by the Timber Research and Development Association (TRADA) for a range of pitches, loadings and coverings. Manufacturers of standard trusses and beams design and test their products, and offer them for the conditions of use for which they are appropriate. Common lattice roof trusses are amenable to direct structural design methods. More advanced forms, lattice grids and decks and complex shells, require computer aided calculation, and model analysis and testing is often necessary in the evolution of the design.

*Imposed loads* on roofs are those due to snow, wind and any moving traffic. The snow load will be appropriate to the geographical location of the building, and to its exposure, from which may be deduced the thickness of snowfall, including any drifting on the roof. Wind loading on roofs varies with the locality and exposure of the building, being more severe adjacent to the sea, at high altitude or in open country, than in sheltered urban locations inland. Lightweight roof structures are especially subject to racking and distortion under wind pressure and suction, and light deckings to stripping. This may be caused by wind pressure under exposed edges, or by suction, which may pull the decking over its fixings, or may carry the fixings and part of the supporting structure with it. Light roofs of low pitch are affected by wind uplift due to suction (see Fig. 4). This can exceed the weight of the roof structure, and provision must be made for anchorage to heavy walls or the equivalent. If the walls are of light panels or cladding, the resistance to uplift must be through columns to heavy foundations. A typical method of restraint is described on pages 45-46.

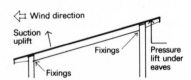

Figure 4. Wind effect on monopitch roof

On pitched roofs or light deckings, allowance for traffic loads may be limited to that necessary for occasional maintenance or repair only. A higher load value may be necessary for pedestrian trafficking on a balcony or terrace. Flat roofs to be used by considerable numbers of spectators will require to be designed for a known number and density of persons.

*Incidental loads* may be imposed by water tanks, suspended ducts, pipework, and plant, such as lifting tackle. The roof structure may be designed to receive anticipated loads at given points, or may be strengthened or modified locally to take known static loads. Steel or timber networks lend themselves to the convenient provision of fixings for suspended equipment, and in lattice grids and decks the load is distributed in several directions. Dense reinforced concrete shells are not so amenable to the making of, or provision for, fixings, or for the hanging of intense local loads.

Within limits, deflection or distortion is not always structurally damaging but may be unsightly, as in the sag of inadequately sized purlins. Deflection is most noticeable when a member is viewed along its length, or when seen parallel to an undistorted line. Such movement may cause rigid finishes like plaster to crack and spall off. Designed roof or gutter falls may be reduced or eliminated, flashings may be split, and roof glazing stressed and cracked. The movement tolerance of a roof structure must be assessed in relation to the susceptibility of its associated components and materials to damage.

## Thermal insulation and movement

A major function of thermal insulation in the building envelope is to contribute towards the achievement of reasonably steady temperatures which the occupants will find tolerable and comfortable for efficient activity, or alternatively a temperature required for a specific process. In general terms, this means a temperature range of 15 to $22°$C (60 to $70°$F), suitable for most living and working purposes. If a space is to be refrigerated, very efficient insulation will be required. At the opposite extreme, the excess heat generated in a steel foundry must be dissipated through single skin ventilated surfaces. If the roof function is to be no more than simple shelter from weather, as in an open barn, insulation is of no value and a thin surface cover is adequate.

Accepting the need for thermal insulation in a roof, its importance will be greatest where the area of roof is a high proportion of the total exposed building surface, as in extensive single-storey factories and schools. The provision of a high standard of thermal insulation has several economic benefits. The capital cost of the heating installation will be lower than that required to compensate for high heat losses, and consumption of fuel will be less, benefiting the owner or occupier financially and conserving home-produced or imported fuels. Thermal insulation also reduces solar heat gain by day, and heat loss by radiation to a clear night sky in winter. If a low density insulant remains uncompacted in use, and a reflective insulant stays bright, functional efficiency is maintained without cost.

The demand for adequate heating in dwellings has led to increased statutory requirements for thermal insulation of new and existing houses. This is normally easy and economical to instal wherever there is reasonable access. The Building Regulations 1965 (Part F) require the roofs (Reg. F3) and walls within roofs (F4) of new dwellings to be insulated to a U-value of 1.41 W/m$^2$ deg C (0.25) for the roof and 1.70 (0.30) for the walls. These values may be achieved by the constructions and insulation given in Tables A and B of Schedule 11, to which Regulation F7 refers. The required minimum levels of insulation are easily achieved, and should preferably be exceeded.

The Thermal Insulation (Industrial Buildings) Act 1957 was introduced to reduce the heavy consumption of fuel in factories; these were commonly clad with single skin materials. The Act has assisted in the achievement of appropriate internal temperatures, reducing high winter heat losses and discomfort from solar heat gain in summer. The stipulated insulation values relate to the required internal temperature, the highest being 1.70 (0.30) with an internal temperature of 15$^\circ$C (60$^\circ$F) which is provided in sedentary activities, reducing with the design temperature.

This legislation led to the introduction of double-layer corrugated deckings, with a quilt sandwiched between an external structural sheet and a supporting sheet lining (Fig. 12l). Another common decking has fibreboard, cork or expanded plastics sheet bonded to the upper surface of a corrugated metal deck and covered by multi-layer bituminous felt or similar weatherproof covering (Fig. 12k). Single-layer decking may be improved in thermal performance by the application of a sprayed asbestos fibre layer about 12 mm thick to the underside. This treatment is useful also for reducing internal reflected sound.

Few thermal insulating materials have a structural spanning function, owing to the poor flexural strength of most low-density materials. The structural grade of plain woodwool slab in 50 mm thickness spans up to 600 mm, but with pressed-steel channels incorporated in the long edges of 50 and 75 mm slabs, the maximum span is increased to about 3 m. Some patterns of channel-reinforced woodwool slab have a rebate over the channel, filled by a cork strip, which prevents a "cold bridge" between interior and exterior

16

through the steel, leading to surface condensation on the channels (Fig. 5b). Others embody tongue and groove channel profiles, to ensure alignment and structural continuity between slab edges. Compressed strawboard in 50 mm thickness has spanning characteristics similar to plain woodwool. Exceeding the recommended spans leads to deflection and an accumulation of shallow pools of water on very flat roofs.

Thick softwood of up to 75 mm is also used as decking, the inherent insulation value of which may be increased by a top layer (Fig. 5a). Solid precast reinforced foamed concrete units are produced for roofs, the entrained gas bubbles providing the insulating function (Fig. 5c).

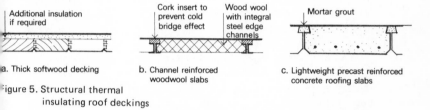

a. Thick softwood decking

b. Channel reinforced woodwool slabs

c. Lightweight precast reinforced concrete roofing slabs

Figure 5. Structural thermal insulating roof deckings

For most other roof forms, thermal insulation is provided by non-structural materials of two types:

1. Those of low thermal conductivity.
2. Those of high thermal reflectivity.

Materials of low thermal conductivity have a high percentage volume of gas or air voids, which retard the transmission of heat. Most efficient are materials with a closed air or gas cell structure, such as the familiar expanded plastics, used in board or granule form, a few millimetres thickness of which give insulation equal to a substantial thickness of brickwork, dense concrete or stone.

Typical of this class of insulator are:

(a) *Quilts* consisting of glass fibre, rock wool or slag wool (classified together as mineral wool).

(b) *Slabs* of woodwool, strawboard, fibreboard, expanded plastics, cork, semi-rigid glass-fibre or foamed glass. Thick low-density softwood strip (preferably 50 mm and above).

(c) *Granulated or nodulated materials used as loose fills,* in layers on ceilings, or to fill cavities. Pelleted slag wool, exfoliated vermiculite (a naturally occurring micaceous material which expands when its contained water is vaporized by heating), are examples.

(d) *Plastics foamed in situ* and injected into cavities to fill them. (Note that the foam stabilizes the insulating air in the cavity by

17

incorporating it as millions of very small cells within the material.)

(e) *Air or gas cells* within a basically high-density material, as in foamed concrete or screed.

(f) *Lightweight-aggregate concretes and screeds*, which, to be effective, must be of adequate thickness, dried out, and kept dry.

(g) *Sprayed insulation*, of asbestos fibre with water-activated binders, or lightweight-aggregate plasters, applied to a thickness of 12 mm or more, on exposed protected internal surfaces.

The presence of moisture in an insulating material will reduce its efficiency, since water, with its high thermal conductivity, displaces a proportion of the air. Porous insulants must therefore be installed dry, or dried out, and protected from water gain from rainfall, other wet processes, interstitial condensation, or roof leakage. Expanded plastics, with closed-cell structure and low vapour permeability, are less affected than open-cell products.

If an insulating layer on or within a roof structure is likely to take up condensation water from humidity within the building, a vapour-impermeable membrane must be provided on the "warm" side of the insulation. (See BRS Digests 8, 51 and 110 (Second Series).) Ventilation of roof spaces also helps to reduce moisture, provided that an upward and outward air flow can be achieved. This may involve the provision of special ridge vents on a double-pitched roof having continuous membrane covering. Although ventilation should preferably be above the roof insulation, a protected perimeter slot of about 6 mm depth to the eaves of a house having strawboard deck-on-timber-joist flat roof would not materially reduce insulation. On the other hand, an aluminium-foil faced plasterboard ceiling would be helpful in increasing the insulation below the ventilation and in reducing water vapour flow, the foil (but *not* the board joints) being impermeable.

Reflective insulants depend for their function upon their ability to reflect incident short-wave radiant heat. This effect may be experienced by feeling the beamed heat from the curved chromium-plated reflector of an electric fire. A double-sided sheet of aluminium foil facing cavities will reflect heat from sources inside and outside the building. To remain effective, the foil must remain bright and free from surface corrosion or dust. White and other near-white colours reflect a high percentage of heat and light, and are used to reduce solar heat gain through roofs, but they do not reduce heat loss.

*Location of insulation: low-density materials*

*Structural insulating decks and slabs* These provide integral insulation. If inadequate, the insulation may be increased by

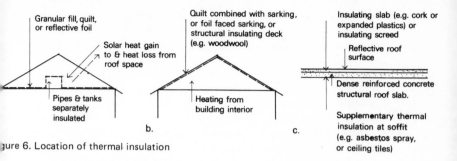

Granular fill, quilt, or reflective foil

Solar heat gain to & heat loss from roof space

Pipes & tanks separately insulated

Quilt combined with sarking, or foil faced sarking, or structural insulating deck (e.g. woodwool)

Heating from building interior

b.

Insulating slab (e.g. cork or expanded plastics) or insulating screed

Reflective roof surface

Dense reinforced concrete structural roof slab.

Supplementary thermal insulation at soffit (e.g. asbestos spray, or ceiling tiles)

c.

Figure 6. Location of thermal insulation

non-structural board insulation above them, or by ceilings below that are themselves insulating, or which support insulating quilt or loose fill (Figs. 5a, b, c and 19a).

*Lightweight slabs* These are normally fully supported by the structural roof, but may be partially supported, as when fibreboard or cork slab is bonded to the top surface of corrugated-metal decking and spans across the corrugations 100 mm or so (Fig. 12k). Slabs or sheets of adequate rigidity will span short distances between joists or beams, as roof deck or ceiling. Woodwool slabs may be used to line shuttering for *in situ* reinforced concrete, remaining on the concrete after the removal of shuttering. Precautions must be taken to ensure permanent and overall bonding, to prevent the detachment of overhead linings when the building is in use.

*Quilts* These are draped over ceiling joists with edges closely butted or lapped. Alternatively they may be supplied pre-cut to narrow widths for fitting between joists. Another use is between the two surfaces of a double-skin asbestos cement or metal deck. A recent introduction is a quilt, faced with sarking felt, laid over pitched-roof rafters to provide underfelt and insulation in one operation. With insulation in this position, plumbing in roof voids is protected against frost while the building is heated, but the additional volume will increase the total heat requirement of the building. Quilts incorporated in cavities in pitched roofs must be secured against sliding down, with consequent loss of insulation (Fig. 6a, b and 19a).

*Loose fills* These are spread evenly to a suitable thickness on a continuous flat surface, such as a ceiling. They are not suitable for sloping surfaces, except where a parallel cavity of 40 to 100 mm is to be filled, or access is available for complete filling, or a prefabricated unit is pre-filled.

*Lightweight screeds* These are laid on the upper surfaces of concrete slabs. On pitched or curved surfaces a constant thickness of 50 to 75 mm is laid. The screed must be dry to be fully effective. (Fig. 6c).

19

*Sprayed insulation* This is applied to the soffits of dense concrete surfaces and will follow any profile accessible to the spraying equipment (Fig. 6c). It may be used on metal lathing, or on sheet and deck materials affording the necessary adhesion. Low density assists in preventing surface condensation, and in reducing reflected sound concentrations where curved hard concrete soffits have acoustic points of focus at the centres of radius of cylindrical and spherical shells. Their stability under heat, their incombustibility and thermal insulating qualities make asbestos and vermiculite sprays useful as applied fire resistant coatings.

Highly efficient insulation should not be used alone on the soffits of dense concrete slabs and shells of large area, particularly if the roof finish is dark in colour. Large thermal movements will occur under climatic temperature changes, aggravated by the effectiveness of the insulation in retarding downward heat loss and thus causing high temperatures in the slab. Top insulation should be provided to reduce the temperature range in the slab, but this insulation will cause high temperatures in the weatherproof covering, which may be detrimental to it. A light-coloured surface will reduce temperatures in the covering. (See *Thermal Movement,* below.)

*Reflective insulating materials* (Fig. 6)
*Plain aluminium foil* Used alone, it is stapled to joists, or to the underside of rafters. A patent double-layer crimped form is available. As the foil is liable to tear, it is also supplied bonded to one or both sides of a stout kraft paper.
*Foil-faced plasterboard* ("insulating plasterboard"). A layer of foil is factory bonded to one side of 9.5 and 12.7 mm plasterboard. Used as a ceiling board, the reflector faces the roof space. A layer of building paper or polythene sheet stapled to the tops of the ceiling joists of pitched roofs reduces dust accumulation.
*Foil-faced reinforced bituminous sarking felt* Used as a normal underfelt on pitched roofs between rafters and tile battens.

As no additional labour operation is required the last two materials provide thermal insulation for the net extra cost of the foil application. Generally speaking, quilts, loose fills and reflectors are lower in cost than boards and slabs of comparable functional efficiency, but the latter are often suitable for roof deck or ceiling use, this dual function making them economical.

Corrugated natural aluminium decking has a useful thermal reflectivity, so long as it remains bright and unweathered.

On roofs, solar reflectors in common use are:
*Tallow limewash* Applied to bituminous surfaces annually in spring. It dries to a dense white surface, and is at its cleanest and most efficient during the following summer.

*Aluminium bituminous paint and compounds*    This dries to a silver finish. Note that oil-based paints, including emulsions, should not be used, as the oils affect bitumen in felt, and also asphalt roofs.

*Single-layer plastics or synthetic-rubber coverings*    These are available finished white.

*White or light-coloured stone chippings*    These are bonded *in situ* to bitumen felt or asphalt, at pitches up to 10 deg.

*White mineral-surfaced bituminous felts*    These are used on roofs pitched above 10 deg.

*Light-coloured asbestos cement tiles*    These are used as paving on flat roofs, bedded and jointed in bitumen. The 25 mm thickness has a low-density insulating core.

While the provision of reflective surfaces on roofs is almost universally desirable on functional grounds, visual considerations may preclude them on visible roofs. A surface that reflects light and heat may cause glare and discomfort in rooms above and adjacent to it. Reflective efficiency depends on retention of a light colour. Deposits from air pollution, leaves, dust, and moss growth on poorly-drained flat roofs will darken the surface. To provide effective water clearance and washing effect, reflective roofs should be well drained.

*Roof glazing*    Single-layer glazing, whether glass, or clear or translucent plastics, will reduce the thermal insulation of most roofs, and solar heat gain will occur beneath the glazing. White and green paints have been formulated for summer application to large areas of roof glazing, as a makeshift to reduce heat discomfort and glare. Areas of glazing should be designed for adequate but not excessive natural light on the working plane. Many factories are constructed with well-insulated roofs without glazing, as this increases cost and creates vulnerable interruptions in the covering. In such situations permanent artificial light has to be used to give even and controlled illumination.

Some roof glazing systems may be obtained as double units. Plastic domelights may have a sealed dry air or gas space between two membranes. In patent glazing, which uses glass sheets 600 mm wide carried on complex structural tee bars, two glass layers may be used. These are divided by a pressed-aluminium spacer channel, giving about 12 mm airspace. Although improved thermal insulation will be obtained at additional cost, the insulation value will still be inferior to a well-insulated, unglazed roof construction.

*Thermal movement*
The wide temperature range experienced in rigid concrete roof slabs of extensive area, between a sunny midsummer day and a clear winter night, will cause significant expansion and contraction.

Slated, tiled or corrugated-sheet roofs take up thermal movement in multiple butt joints, sliding overlaps, and by flexing of corrugations. Network roof-structures lying beneath the covering, experience the temperature conditions of the interior. In buildings having flexible structural materials and finishes, which are therefore tolerant of movement, movement may not be visible or damaging.

The movement of dense-concrete slab roofs is particularly marked in the case of shells, whose thin membranes gain and lose heat rapidly. A folded or curved concrete surface will take the movement by a rise and fall in the curved profiles, but in any straight length, extensional movement will occur. The worst conditions apply when the slab has a dark coloured external finish, and carries the insulation entirely beneath it. Rigid concrete slab and membrane roofs should be efficiently insulated *above* the slab, and if possible should have a reflective outer surface. This treatment should be associated with the subdivision of the roof by expansion joints, into areas in which the thermal movement is so reduced and limited as not to cause cracking and damage of susceptible materials and finishes.

# Chapter 2

## Sound insulation

For most roofs, minimum deadweight is an important criterion of design. Light insulated decks, or thin membranes, are therefore almost universal on long spans. This requirement is in conflict with the need for heavy mass for the reduction of sound transmission. In many long-span factories, stadia and hangars, *internal* noise levels are so high that reduction of *external* noise is unnecessary. On the other hand adequate sound reduction through the roofs of auditoria and concert halls is of paramount importance, and they are increasingly affected by overhead noise from aircraft. An outer concrete membrane of adequate thickness may be associated with an inner, structurally isolated shell, or a heavy ceiling suspended from hangers of minimum cross section, to provide sound reduction. The depth of the interspace and its lining with acoustic absorbents can thus be important factors.

Apart from having low airborne sound reduction, lightweight roofs also transmit impact noise from the drumming of rain and hail. While the severe noise thus generated on exposed corrugated metal decks may not obtrude above the general noise level of a factory, it can be of considerable nuisance in domestic buildings, as occupants of the early post-war aluminium and asbestos-cement clad prefabricated houses experienced. The noise of heavy rain is clearly audible beneath roofs of bituminous felt laid on woodwool or strawboard deck on exposed joists, although not normally heard distinctly beneath a tiled roof with plastered ceiling.

Reasonably high airborne sound reduction is best achieved in short and medium spans by the use of concrete slab roofs having a dead weight of about $250 \text{ kg/m}^2$ or more, with the addition of screed and a plastered or suspended ceiling. Any form of roof light will lessen the reduction achieved by a heavy construction in excess of 40 to 45 dB. Double roof glazing with a deep, sealed airspace of 100 mm or more will be required to achieve 40 dB insulation. Glazing is best omitted from a roof whenever a really high reduction of overhead sound is important.

Good sound insulation of a roof may be important where it covers a noisy industrial process, in order to avoid nuisance to surrounding buildings. Where very loud noises are created, a heavy sealed enclosing envelope may be indicated, coupled with physical separation from nearby buildings.

The Building Regulations 1965 stipulate requirements for sound insulation of the walls and floors separating dwellings. There is no requirement for domestic roofs, as these do not occur between separate occupancies.

## Fire protection

Roofs may be ignited externally by exposure to intense radiant heat and flame from fires in adjacent buildings, or alternatively by burning material carried on rising convected hot air, which may travel considerable distances to lodge on roof coverings. Fire will thus gain a hold and spread if combustible material in the covering or its support catches fire. Thatch is a definite hazard. Wood shingles or some bituminous felts may also be combustible. Combustible boarding or decking beneath the covering will also contribute to the rapid spread of fire.

Fires occurring *within* the building will readily ignite roof structures of small-section timber, and will distort exposed and unprotected steel trusses and frames, causing severe mechanical damage. Heavy solid-timber sections will remain intact and stable for fairly long periods, as the rate of surface combustion is slow, and a carbonaceous layer is formed, retarding further burning. Precast or *in situ* reinforced concrete structural members are incombustible, and providing that the thickness of section is adequate will remain stable under fire exposure. Such members can achieve statutory periods of fire resistance.

Most lightweight deckings are not resistant to fire for long periods. Asbestos cement shatters; metal distorts or melts. Woodwool ignites slowly, as does compressed strawboard, particularly if the latter is asbestos-paper faced, retarding the surface spread of flame. Fibreboard and untreated thin boarding are definite hazards, particularly if continuous cavities occur behind such linings, permitting fires to progress in inaccessible positions with a free access of oxygen. Cavities should therefore be avoided or "fire-stopped" at intervals, to prevent access of air, and the cavity from functioning as a flue.

Structural frameworks may be underdrawn with fire-resistant linings. If exposed, network structures may be coated with asbestos spray, although this is a laborious operation, particularly on complex grids. The spray may also be used on non-fire-resistant decks, or these may be closely underdrawn with low-density asbestos insulation board (not to be confused with asbestos-cement sheet).

Roof linings should have low surface flame spread rate, preferably Class 1 of BSS 476, desirably without surface retardants or impregnation with fire-retardant chemicals.

Fires occurring in large single compartments, such as factory production areas, spread laterally, and the interior may become smoke-logged, making it difficult for firemen to locate, approach and suppress the outbreak. The roof space may be subdivided by incombustible "curtains", each compartment being provided with automatic vents to vent smoke and hot gases at the seat of the fire, thus localizing it, and keeping the surroundings clear of smoke.

The Building Regulations 1965 contain no requirements for periods of fire resistance of the structural elements of the roof. Where, however, high fire risks occur from hazardous industrial processes or storage of inflammable contents, or when a manufacturer has a bad record of industrial fires, fire resistance of the whole roof may be made a condition of fire insurance. The Regulations stipulate spread of flame characteristics of the roof soffit (E14) and the designation of the roof finish and its underlying construction in respect of external fire exposure (E15), which relates the finish and construction to the volume of the building, its use, and the proximity of the roof to adjacent boundaries and buildings. In effect, roofs whose combustibility makes them most liable to fire from external ignition, or which, when burning, may ignite neighbouring roofs, are required to be used only on buildings of limited volume, of low-hazard occupancy, and at specified distances from the site boundary.

Roof coverings and supporting constructions are allocated, in Schedule 10 to Regulation E1, a two-letter designation, derived from tests specified in BS 476: Part 3 "External Fire Exposure Roof Tests". The first letter relates to fire penetration, the second to flame spread over the external surface. The letters are A, B, C and D, in combination, A being the highest rating, D the lowest. To give examples of extremes, AA is achieved by non-combustible coverings, in most cases even when used over combustible construction, and lower designations, such as CC, are allocated to combinations of combustible constructions and coverings, such as bitumen-felt strip slates on timber boarding, in pitched roofs. Constructions of the more hazardous designations are restricted to smaller buildings, sometimes of specified uses, and at substantial distances from site boundaries.

It should be noticed that the combustible material bitumen is least hazardous when combined with incombustible materials, as in asbestos fibre-based felt, or as mastic asphalt, which contains a high proportion of mineral filler; and also when closely in contact with non-combustible and fire-resisting deckings or slabs.

# Adaptability and modification

Structural roof systems differ widely in their requirements for support. A traditional reinforced concrete slab, or timber-joisted construction may need continuous support at the building perimeter and intermediately by walls or beams. In contrast a reinforced concrete cylindrical shell (Figs. 2, 29) or steel space-deck (Figs. 21c, e) can be carried by columns at widely-spaced centres. Such roofs allow maximum freedom in plan subdivision and in the selection of walling materials within the covered area.

The *in situ* concrete roof will not readily lend itself to liberal perforation of its surface or beams, but the network structure provides visible zones in which pipes and ducts may be located, limited only by the positions of essential structural members, and by any loading restrictions. Solid concrete constructions inhibit adjustments to services and are usually perforated with sleeves or holes, during construction, in structurally non-critical positions. Precast concrete members are either perforated during manufacture, or provided with regularly spaced holes for unknown future requirements.

Although timber beams and solid or plywood web or box members are capable of being holed easily for penetration, the designing engineer or manufacturer must be consulted about the location of holes and limits of cutting away.

A simple traditional timber raftered roof requires the perimeter support of rafter feet, and the intermediate support of dividing purlins. This is normally provided by loadbearing walling. Removal or modification of such walling is difficult (Figs. 8b, c). Deep trussed purlins (Fig. 8a) are often used to span between party walls, or between columns or piers, and can carry the whole rafter and ceiling joist system, thus enabling bounding and dividing walls to be located freely, under the roof cover, restricted only by the supports of the purlin ends. Wide variation of plan form is possible, and internal subdivision is not necessarily static for the life of the building.

Metal and timber lattice decks offer the great advantage of long spans between supports, permitting free internal planning. Decks of this type, based upon a standard unit (Figs. 21a-e) may be dismantled and re-assembled to a different plan. Although it does not appear that this facility has been exploited to any great extent, a roof system that can be altered or extended, using existing or new standard parts, commends itself for buildings that may require modification at frequent intervals.

*In situ* surface structures, and precast structures grouted *in situ* for continuity, are most suitable for buildings not requiring modification of support location, internal height or structural cover.

In general terms, alterations to buildings involve adding external units, or plan modification within defining floors and roofs. Companies specializing in the lifting of structural floor slabs at the time of construction by co-ordinated jacking, have successfully lifted existing roofs, inserting additional column lengths. Facilities for this technique might be embodied initially in buildings that may require increased headroom, or the introduction of additional floors at some future date.

## Maintenance, demolition and renewal

The useful life of a building is ended by demolition or dereliction, often following a period of obsolescence and decay. External enclosing elements are subject to differing internal and external atmospheric exposure. Many economic walling materials have an inherent durability, coupled with a low maintenance requirement, which exceeds the useful life of the building. Traditional pitched roof coverings of slate or tile may be expected to remain sound, if correctly selected, fixed and maintained, for the life of a house, say 60 years. Failure often affects only a few of the units, easily replaced if still available, unless faulty design and installation leads to large-scale stripping. It is useful for the designer to specify that on completion, a number of the roofing units are stored within the building, as a safeguard against future lack of availability.

If the roof structure is in dry non-corrosive conditions, or protected effectively from insect or fungal attack, or an aggressive atmosphere, it will be preserved for an extended period, perhaps beyond the required life of the building. Industrial structures are envelopes protecting the productive activity within. In certain industries, expansion is rapid and modification of processes frequent. The enclosure may inhibit change, even if originally designed for adaptability. Removal and rebuilding may be required frequently, even after a decade or so. The designer's initial problem, therefore, is to produce a structure which is economical in first cost, while at the same time providing adequate environmental standards and reasonable maintenance in relation to materials selected for adequate life expectancy. Provided that their removal is easy, materials offering long life with economy are obviously useful.

In normal non-aggressive atmospheres, lightweight steel, timber latticed, network structures or portal and purlin forms commend themselves. This is especially so when they are associated with insulated deckings of galvanized or plastics coated steel, aluminium alloy, asbestos-cement or reinforced woodwool, protected by bituminous felt, plastics or synthetic rubber sheet roofing. Precast reinforced concrete frame and purlin superstructures are inherently durable and require no maintenance if self-finished. They may be re-usable if no large scale *in situ* grouting for continuity and stability

27

is employed. For rapid clearance, or for re-use, components should dismantle easily. However, some industrial renewal programmes are so urgent that summary demolition, without regard for component recovery, is vital. In such cases, the materials will have only scrap value.

Monolithic and continuous *in situ* reinforced concrete roof structures, including shells, are most suitable for enclosing static processes, for which their durability is often of advantage. Their demolition involves the use, often at elevated working levels, of pneumatic breaking equipment and high-temperature oxygen flame cutters. The resulting materials are useful only as hardcore or filling.

The main structural members of roofs are sometimes exposed, in the form of upstand solid or lattice beams, or suspension cables and masts. The roof slab is underslung, and has an unimpeded soffit. The use of inherently durable structural materials, or durable coatings, is essential if deterioration is to be avoided. Points of suspension of cables, or junctions between roof covering and beams or trusses, present problems of water sealing. Suspension cable roofs are often used on large exhibition pavilions, providing very large and uninterrupted internal spaces.

## Economic factors

It is not possible here to give specific information about the initial or subsequent maintenance costs of a roof system in relation to comparable alternatives; or to tabulate the factors which affect the cost of roof structure. Generally, a structure is used over the range of spans for which it has been found to be economical, in relation to other forms, and to the cost of the roof element in an overall building cost plan. The use of a system at the upper limit of its span capability normally results in higher cost per unit area of plan covered than at shorter spans. The covering of large plan areas without internal support by walls or columns results in a greater weight of structural material in relation to live loads on the roof, and in higher costs of fabrication, transportation, assembly and erection. The cost of shuttering a single large span *in situ* reinforced concrete shell will be high in relation to total expenditure on the roof, as there may be little re-use of shutter materials, which do not remain as part of the completed structure.

Certain classes of buildings require unobstructed floors or internal spaces, but in general terms, spans should be the lowest commensurate with the efficient functioning of the building during its life, insofar as its use may be determined. In addition to structure and covering, costs will have to cover supplementary materials necessary for thermal insulation, acoustic control, resistance to fire and economic maintenance, wherever these criteria are important. A cost comparison between alternatives will be on a basis of their ability to satisfy them.

While the tenet that *increase of span increases cost* still largely applies in structural design, the introduction of new roof systems and new design techniques are tending to increase economic spans. Such systems offer reductions in the overall dead weight of loadbearing and covering elements. These improvements have occurred mainly in the three principal materials, steel, timber and concrete. Aluminium alloys are employed for long-span structures in which deadweight is a critical factor. High material and fabricating costs tend to inhibit the use of aluminium for lower spans, but mechanical connectors for tubular aluminium grids are being developed. Reinforced plastics are also used in structural roofs, mainly in stressed-skin units. In this connection the integral waterproofing characteristics of structural plastics and the possibility of varying light transmission through them are attractive.

The selection of roof type is normally based to a great extent on cost comparisons. Occasionally, however, considerations of prestige and maximum architectural quality call for a structural and constructional virtuosity in the roof form, for which the high cost cannot be justified in purely functional terms.

The comparative cost study will be most critical where the area of roof forms a high proportion of the total external surface, as in single-storey factories, but less so where the area of roof is small, as in tower blocks of small plan area.

Where a roof is economic or practicable for a certain range of spans, or where cost and structural efficiency are affected by design factors such as rise to span ratio, or repetition of identical units, these factors will be given in the text.

# SECTION TWO
## Skeleton roof-structures

This Section is mainly concerned with the many forms of skeleton roof-structure currently in use, and its three Chapters deal with *Pitched roofs*, *Flat roofs* and *Three-dimensional structures* respectively

# Chapter 3—Pitched roofs

## Timber pitched roofs

Timber pitched roof structures may be divided into two broad classifications. The first is constructed entirely *in situ* from basic timber sections using simple jointing methods. The second utilises trusses, trussed rafters and trussed purlins which are assembled on site before being lifted into position, or are completely fabricated at works. The first category is based upon traditional methods with modern refinements; the second upon calculated and tested designs produced by the Timber Research and Development Association (TRADA), by the makers of connectors and nail plates, and by timber engineering companies, who market roof components and use new jointing techniques and rationalized production methods. These manufacturers can supply trussed rafters for many combinations of pitch, profile and span, from single house contracts upwards. Cost per unit is, of course, lowest where large scale repetition is required. *The traditional timber pitched roof* is based upon the utilization of standard sawn or commercially surfaced (planed) rectangular softwood sections, of sizes to suit load and span conditions. These are ordered and delivered to site in appropriate lengths, cut and assembled in position, located and fixed mainly by nailing. All cutting, jointing and assembly has to be done by carpenters at roof level. Covering in of the building interior must await completion of roof carcassing, as there is no prefabrication to shorten construction time. Trussed rafters, however, may be assembled at ground level, or supplied complete from the factory, thus reducing site work time before covering.

A typical pitched roof covers a rectangular plan, a series of connected rectangles, or a regular polygon. A sheet roof covering allows a tapered or irregular straight-sided plan to be covered, using rafters at constant pitch to avoid warping the roof plane, and using a falling gutter or ridge line. To avoid unsightly cutting, parallel ridge and gutter lines are desirable for slates and tiles, trimming being restricted to hips and valleys.

Using basic timber sections, many variations of traditional pitched roof may be designed, detailed and constructed. The pitch will be determined by the requirements of the covering. Plain tiles

33

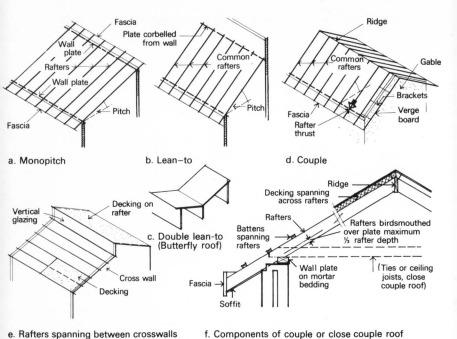

a. Monopitch

b. Lean-to

d. Couple

c. Double lean-to (Butterfly roof)

e. Rafters spanning between crosswalls

f. Components of couple or close couple roof

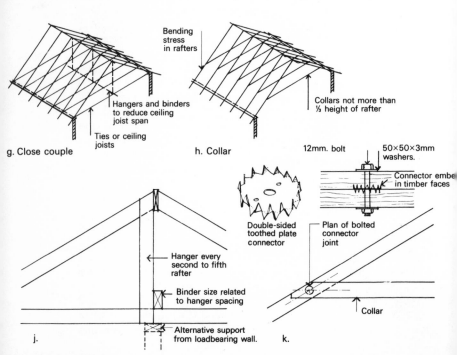

g. Close couple

h. Collar

j.

k.

Figure 7. SINGLE ROOFS

require at least 40 deg, mineral surfaced multi-layer bituminous felt 10 deg. Below this pitch the roof may be regarded as flat.

Pitch may also be determined by any requirement for roof-space utilization, for which the pitch will be steep. Alternatively, there are combinations of steep and shallower pitch, as in the mansard roof. Pitches above the minimum for a covering are, of course, permissible, having due regard for any special fixings at steeper than normal inclination.

Low-pitch trussed rafter roofs for housing give low headroom in and poor accessibility to roof storage space. Steeply pitched strutted-purlin roofs are better for this purpose, but if substantial use for storage is anticipated, ceiling joists should be made larger than the minimal 75 by 50 mm or 100 by 38 mm, if ceiling cracking due to joist deflection is to be avoided.

When timber was plentiful and low in cost, roof members were made of ample size in good-quality building work. Restrictions on the importation of softwoods after the Second World War led to the publication of official *Economy Memoranda for the Use of Timber in Building*. Sizes and spacings were given for structural members in floors and roofs, related to the position, loading, and the timber quality. Tables of minimum sizes of members in pitched roofs permitted by the Building Regulations 1965 are given in Schedule 6 of Regulation D14 (Tables 2, 7, 8, 9 and 11). For economy in domestic work, rafter sizes are usually restricted to 125 by 50 mm or less by the use of purlins, ceiling joists to 100 by 50 mm using binders, and purlins to about 225 by 75 mm by the introduction of supports at suitable centres.

Softwoods commonly used in timber roof construction are the carcassing grades of sawn European Redwood or Whitewood of Swedish, Finnish, Russian or Polish origin, Douglas Fir, with dimensional variations allowed by British Standard 1860, Part 1 and CLS (Canadian Lumber Standards) and Pacific Coast Hemlock, which is supplied planed all round with arrises slightly radiused. This is accurate in size, and easy to handle.

*The lean-to and monopitch* are the simplest timber-pitched roofs (Figs. 7a, b), usually limited to 3 to 4 m span. Rafters of the lean-to are supported on a wall-plate directly carried on the wall at the lower level, and a plate corbelled from the wall at the upper level. The monopitch rafter bears on to plates at two levels, and is therefore simply an inclined plane.

The wall-plate, a common feature of timber roofs, is a softwood section bedded level on the top of a wall in cement-lime-sand mortar. Its functions are to distribute the loads from rafter feet evenly over the wall, to provide a level bearing, and a location and fixing for rafters and ceiling joists (Fig. 7f).

Rafters and joists are skew (diagonally) nailed on to the plate with flat-headed round wire nails. The traditional wall-plate size of 100 by 75 mm was reduced to 75 by 50 mm after the war, but plates are now usually 100 by 50 mm. The wall-plate may be carried on or replaced by timber beams carried on posts, or by a timber lintel spanning openings at rafter level.

Rafters are birdsmouthed (notched) over the plate, to a maximum of one third of their depth. This provides a dead bearing on the plate and permits alignment of the top surfaces of the rafter. Rafter spacing varies from 400 mm to 610 mm centres, dependent upon loading and the covering material. Sawn softwood battens that carry slates or tiles act as small "beams", collecting the load of the covering and transmitting it to the rafters. They are 38 by 19 mm or 25 by 25 mm for rafters at 400 to 450 mm centres, but must be increased to 38 by 25 mm or more to span 610 mm, otherwise there will be batten sag, or a springiness that prevents tiling or slating nails from being driven in. Plasterboard of 9.5 mm spans 450 mm maximum, and 12.7 mm thick board must be used for ceiling joists at 450 to 610 mm centres.

Sheet materials used as decking on pitched roof rafters require very accurate rafter spacing, and rafters exposed at the eaves must also be very evenly spaced for good appearance. When bolted connectored trussed rafters using twin rafters are used on such a roof, one rafter is cut off at the plate, and the other left projecting. If the eaves are flush, or have a soffit, and battens are used, rafter spacing is less critical, maximum centres depending upon tile load and batten size.

*The butterfly roof* (Fig. 7c) is a variant of the monopitch, two roof planes meeting to form a valley gutter at the lowest level. The lining of a valley gutter to a tiled or slated roof is very prone to blockage and leaks. This type of roof is best restricted to those covered with jointless or sealed sheet materials, and should be used only in areas clear of trees. Rafters are supported at valley level by a beam or loadbearing wall, which carries the wall-plate.

*The couple roof* (Figs. 7d, f) has rafters bearing on wall-plates at the eaves and meeting at and nailed to a ridge board at the apex of 19 to 32 mm nominal thickness. The ridge board may be quite thin, as the rafters should meet in opposite alignment on each side of it. It is a member upon which the rafters are located, and thus prevents lateral displacement. The rafter couple exerts considerable thrust on its support walls, especially if the covering is heavy or there is a snow load. The span of the rafters is limited to about 3 m, above which substantial rafter depth is required to prevent deflection, and the walls may need buttressing to resist outward thrust.

Specialised couple roofs of up to about 18 m span, with steeply-pitched roofs springing from ground level are often used in modern churches. For this purpose, steel, precast concrete or laminated timber beams are used at 2 to 4 m centres. Thrust at ground level is taken by inclined foundations, or by a tie joining the feet of the inverted V beams, below ground slab level. Thrust may also be countered by carrying the rafters or beams on a loadbearing ridge beam, spanning between columns or loadbearing walls.

*The close-couple roof* (Figs. 7f, g, h). In this roof, the thrust is resisted by providing ties in tension at wall-plate level. Rafter spans are up to 5 m but again, rafter size is critical, as the length between plate and ridge is unsupported and not subdivided. The ties may link the feet of each rafter pair, forming ceiling joists, or may be used for a tying function only, at every second or third pair. Spread is prevented by secure nailing at the junction of rafter, tie and plate. The plate-to-plate span of the ceiling ties may be reduced by the introduction of intermediate loadbearing partition walls at right angles to the ties, or by a binder — a timber section which carries the ceiling joists underslung, and is itself carried by hangers, fixed to the rafters at the ridge junction (Figs. 7f, g).

*The collar roof* (Fig. 7h). The roof void, and therefore the total building volume, may be reduced by raising the ties and so forming a *collar roof*. The collars, as the ties are called in this application, are placed not more than one-third up the rafter. Simple direct nailing between rafter and collar will not resist rafter thrust adequately. The traditional joint here is a form of halved dovetail joint. The collar end is a single-sided shallow dovetail, housed 12 mm into the rafter, but as current practice is to replace laborious and complex on-site hand-made carpentry joints by metal fittings, a bolted joint enclosing a double-sided toothed plate connector (Fig. 7k) will be used to resist thrust. The rafter between collar and plate is subject to considerable bending and will probably be 125 or 150 by 50 mm depending on collar-to-plate length and roof load.

The Building Regulation permitting habitable rooms to occupy roof space above rafter bearing level is K8(1). A *scissors truss* (Fig. 10j) also provides room space above bearing level.

A double pitched roof without ceiling joists or ties is constructed by using joists spanning cross walls at up to 5 m centres. Pitch is usually low and may vary between the two planes, and between these planes vertical glazing may be introduced to light internal areas (Fig. 7e). Covering is usually a flexible sealed-sheet system, or metal, on a sheet deck on joists at about 610 mm centres (slates and tiles require battens crossing rafters that follow the roof slope). If the decking is a slab of high thermal insulating value, the joists may be exposed internally, and will be planed to identical cross section. If enclosed by a ceiling, the joists should be treated with preservative, and the void spaces ventilated.

*Purlin roofs* (Figs. 8a-f). A purlin is a beam, used at right angles to the pitch of the roof to carry deckings such as corrugated metal or asbestos-cement, at centres appropriate to the decking span limits (Figs. 12a-f). In pitched timber roofs, the purlin subdivides the rafter span into equal lengths, so that the rafters may be economically sized. Most commonly used rafters are 75 by 50 mm, 100 by 38 mm and 100 by 50 mm, dependent upon load and purlin spacing.

Purlins may be supported at their ends or intermediately by:

(a) *Loadbearing gable, party or partition walling.* Hangers and binders may be used to suspend the ceiling from the purlin (Figs. 8a, d). In loadbearing crosswall terrace housing with light non-loadbearing partitioning, the span of 6 to 8 m would be excessive for a solid softwood section and a deep purlin will conveniently function here as both purlin and binder. It may be of lattice timber truss form, a deep proprietary plywood web beam (Figs. 8a e), or a precast reinforced concrete rectangular section. The rafters and ceiling joists may cantilever from the purlin to the external wall. The wall plate is then eliminated, and the non-loadbearing solid or glazed external wall can be carried up to ceiling joist level. The whole roof framing may be constructed on the ground and lifted into position by crane, using four lifting points on the purlins. This technique demands large scale repetition to justify the use of the crane. After covering further internal work can then be carried out under protected conditions.

(b) *Strutting from loadbearing walls,* using square-section struts bearing on a wall-plate (Figs. 8b, c, f). The wall may be parallel, or at an angle to the purlin, and the strut itself should be at a steep angle. Rafters and purlins may be carried directly on walling taken up into the roof space, but it is preferable to strut from ceiling level, thus enabling the carpenter to construct the roof carcassing complete from plate level. Gable and party walls must be taken up and formed to the rafter line, receiving purlin ends on masonry corbels or mild steel shoes. When strutted, the purlin is notched or housed over the strut, to give a dead bearing.

(c) *Roof trusses, beams, arches or frames,* at right angles to the purlins (Figs. 10, 12, 14).

*Trussed-rafter roofs* (Figs. 8, 9). If the rafter of the truss is on the same line as the common rafters (the purlin, if used, passing

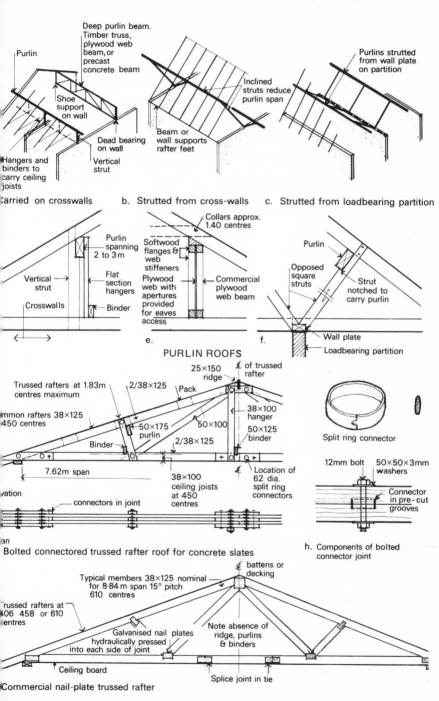

**a.** Carried on crosswalls    **b.** Strutted from cross-walls    **c.** Strutted from loadbearing partition

Purlin

Deep purlin beam. Timber truss, plywood web beam, or precast concrete beam

Shoe support on wall

Dead bearing on wall

Vertical strut

Hangers and binders to carry ceiling joists

Inclined struts reduce purlin span

Beam or wall supports rafter feet

Purlins strutted from wall plate on partition

Vertical strut

Crosswalls

Binder

Purlin spanning 2 to 3 m

Flat section hangers

**e.**

Collars approx. 1.40 centres

Softwood flanges & web stiffeners

Plywood web with apertures provided for eaves access

Commercial plywood web beam

Purlin

Opposed square struts

Strut notched to carry purlin

Wall plate

Loadbearing partition

**f.**

## PURLIN ROOFS

Trussed rafters at 1.83m centres maximum

Common rafters 38×125 450 centres

Binder

7.62m span

Elevation

connectors in joint

Plan

25×150 ridge

₵ of trussed rafter

2/38×125

Pack

38×100 hanger

50×175 purlin

50×100

50×125 binder

2/38×125

38×100 ceiling joists at 450 centres

Location of 62 dia. split ring connectors

Split ring connector

12mm bolt   50×50×3mm washers

Connector in pre-cut grooves

**g.** Bolted connectored trussed rafter roof for concrete slates

**h.** Components of bolted connector joint

Typical members 38×125 nominal for 8·84 m span 15° pitch 610 centres

₵ battens or decking

Trussed rafters at 406 458 or 610 centres

Galvanised nail plates hydraulically pressed into each side of joint

Note absence of ridge, purlins & binders

Ceiling board

Splice joint in tie

**j.** Commercial nail-plate trussed rafter

## TYPICAL 'W' (FINK) TRUSSED RAFTERS

Figure 8a-j

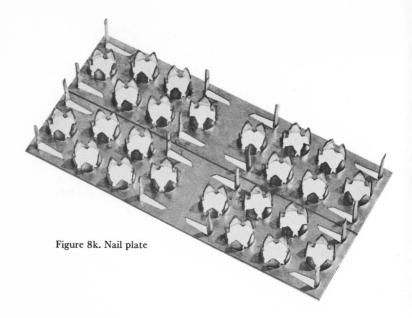

Figure 8k. Nail plate

beneath the rafter and being carried by a notched strut) then a *trussed rafter roof* is created. Trussed rafters in short-span roofs are of two types:

(i) Those based upon *bolted connectored joints*. This type is used at about 2 m centres, that is, at every third or fourth rafter. The purlin, spanning only 2 m maximum, is quite light in section, about 150 by 50 mm to 200 by 50 mm. These trussed rafters may be constructed by site carpenters at ground level and hoisted into position, or they may be supplied by timber engineering companies, part dismantled and folded for transportation. These firms may also supply all the roof timbers.

(ii) Light trussed rafters, with joints between members made with glued plywood gussets, or nail-plates. They are usually placed at 610 mm centres, and eliminate purlins, hangers and binders. Like the trussed purlin roof, the trussed rafter roof may be assembled into sections of suitable size before lifting on to walls (Fig. 9a).

Most trussed rafter roofs span between external walls, and the interior may thus be freely subdivided, but the walls must carry the rafter feet, those of the trussed rafters carrying additional purlin loads.

*The bolted connectored trussed rafter* (Fig. 8g). A nailed face-to-face timber joint has a very limited ability to transmit stress, owing to

40

low strength in shear of the nails, and to the crushing of wood fibres around the small diameter nail shank. A bolted joint is better, if the bolts are of large diameter, and are tightened, using large washers under bolt head and nut, to produce frictional resistance at the interfaces of the members in the joint. However, fibre crushing at the bolt shank may occur.

The design and fabrication of efficient joints in timber structures has been made possible by the development of the timber connector, the glued (or glued and nailed) plywood gusset, and the nail-plate.

The connectors used in timber engineering structures are mainly of two types, the *toothed plate*, and the *split ring*, both made in zinc-coated steel. The toothed plate (Fig. 7k) is a square or circular disc of sheet steel with its edges turned over at 90 deg to form triangular teeth. The connector is holed to pass a bolt. Most commonly used is the double-sided connector, joining two timber members. The single-sided connector joins timber and steel members, the latter being normally a steel flat section used as a tensile member in truss design. A connector is placed over a bolt, between the adjacent faces of a timber joint and the bolt is tightened by a special tool against 3 mm large area, steel washers under bolt and nut. The connector teeth are embedded in the timber faces, and the working stress in the joint is transmitted by the large area of the teeth.

A greater area of steel to transmit load is provided by the split-ring connector (Fig. 8h). One half of the ring fits closely into a circular groove, cut with a rotary cutter, in each face of the joint. Load is transmitted through the area of ring surface in contact with timber in the groove.

In a typical short span trussed rafter roof (Fig. 8g), the purlin is notched into a strut and carries common rafters in the same plane as the trussed rafter itself. Binders, carrying the ceiling joists underslung, are adjacent to the struts and are carried on the ceiling tie of the truss. In the design illustrated three binders divide the ceiling joists into four bays, and the purlin halves the overall span of the common rafter.

Several salient points of timber-connectored truss design may be observed. The strut and tie layout is governed by purlin spacing, which in turn is conditioned by rafter or decking span. The purlins may be notched into a tie *under* the rafter (trussed rafter) (Fig. 8g) or carried *on* the rafter, supported by an extension of an adjacent tie (truss) (Fig. 12d).

Twin rafters and bottom ties are common, using small-section timber. In direct tension, ties may be as thin as 25 mm. Twin compression members have timber packings to stiffen them. Although a truss joint may have only two members, with a single connector, a junction between two twin members in the same plane requires packings and cover plates, and multiple connectors.

To facilitate manufacture and transportation, trusses are usually designed in two halves. This allows the bottom tie to be cut from standard commercial lengths of timber. The truss should be symmetrical in plan longitudinally, without twist and distortion in its members.

Large timber-connectored structures should always be designed by an engineer experienced in this field. Standard designs for short and medium span triangulated roof trusses and trussed rafters are published by the Timber Research and Development Association for various spans, pitches and loadings, and manufacturers of concrete slates intended for use at low pitch also issue standard designs for use with their products. The user of these standard structural designs should not exceed recommended span or load limits, or disfiguring deflections may occur.

*The nail-plate trussed rafter* (Fig. 8j). The recent development of the nail-plate has led to large-scale manufacture of trussed rafters by a number of timber engineering firms. The galvanized steel plates are rectangular, of various sizes, and are of two types. In one, multiple "nails" are pressed out of the plate at right angles to it (Fig. 8k). These are forced into the flush surfaces of members in a joint by roller or hydraulic pressure. It is essentially a factory production technique, requiring heavy tools and jigging, and using stress-graded softwood sections of identical thickness.

The second type of nail-plate is also rectangular, but is perforated uniformly with a rectangular grid of holes. These holes receive close-fitting galvanized nails driven into the timber members of the joint. This technique does not require factory presses and is suitable for the site fabrication of trussed rafters in small quantities. Joint efficiency is dependent on the use of the correct number, gauge and length of nails.

Both types of plate are applied as "patches" on both sides of the junctions of single timber sections of identical thickness in the same plane. This trussed rafter has no projections and can therefore be very compactly stacked for storage and carriage. The main unit made is a symmetrical pitched W trussed rafter (Fig. 8j) available in span increments of 50 mm up to 11.58 m maximum at 17½ deg pitch, and up to 5.38 m at 45 deg. Increments of pitch are 2½ deg. Figure 10 shows the many variations of truss profile that may be obtained. The eaves projection of the rafters is made to the buyer's requirements, and all additional timber members needed to complete the roof framing may be obtained from the one source. Prefabricated gable "ladders" (Fig. 9b) are also supplied, to simplify bracketing for a projecting verge at gable ends. The nail-plate trussed rafter system, using identical units at 450 to 610 mm centres, spans between external walls and provides common rafters and ceiling joists. No purlins, binders or ridge board are required, and the trussed rafters are held in position by a temporary batten, until they are stabilized by final battening or decking.

*The glued plywood gusset*  This is less commonly used in Britain than in Canada and the United States, although one major manufacturer here uses glued 7 mm plywood gussets instead of nail-plates. Low-pitch short-span trusses may have twin plywood strips that act as combined tie and "gusset", between rafter and ceiling tie. In larger trusses, gussets are interleaved with laminated truss members (Fig. 12h). When applied externally, the glued joint may set under pressure, or nailing can be used to apply cramping.

*Openings in timber roof structures*  Openings for chimneys, rooflights or trapdoors may be made, of width up to the distance between adjacent rafters and joists, by the use of trimmers (Fig. 9c). The load transmitted by the trimmers to their adjacent members will be low. However, if two or more members are trimmed out, the trimmer and trimming sections should be increased in width by half, but for simplicity, doubled members spiked together are more commonly used. If a long dormer is to be formed in the roof, the trimmer carries the load of several trimmed rafters, transmitting it to the trimming rafters. Trimmer and trimming members will increase in size substantially, and will probably project below rafter level. Traditionally, hand made tusk tenon joints are made between the members in a trimmed opening, but a range of patent zinc-coated pressed-steel framing anchors is now available for making sound nailed joints, using plain sawn square ends (Fig. 9d).

Figure 9a. Assembled nail plate trussed rafter roof being lifted into place

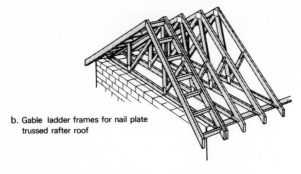

b. Gable ladder frames for nail plate trussed rafter roof

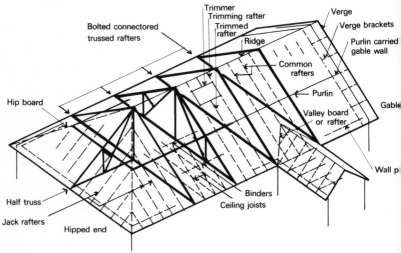

Bolted connectored trussed rafters

Trimmer
Trimming rafter
Trimmed rafter
Ridge

Verge
Verge brackets
Purlin carried gable wall

Common rafters

Purlin

Hip board

Valley board or rafter

Gable

Wall p

Half truss

Binders
Ceiling joists

Jack rafters

Hipped end

c. Diagram of elements of carpentry in pitched roof spanning between external walls

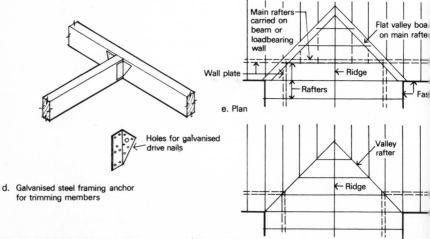

Holes for galvanised drive nails

d. Galvanised steel framing anchor for trimming members

Main rafters carried on beam or loadbearing wall

Flat valley boa on main rafte

Wall plate

Ridge

Rafters

Fas

e. Plan

Valley rafter

Ridge

Figure 9. TIMBER ROOF DETAIL

f. Plan valleys

*Gables, hips and valleys* In the simplest pitched roof employing rafters or trussed rafters, the walling is carried up to the profile and pitch of the rafter top, to form a gable, and the roofing material oversails the wall. A projecting verge is supported by rectangular section brackets carried upon the gable wall and its flanking rafter and projecting to receive the verge board or "barge board". The gable ladder referred to earlier is a prefabricated version of this arrangement (Figs. 9b, c).

The volume of the roof void, and the amount of timber used, may be reduced, and gable walling eliminated by forming a hipped roof. The hip rafter resembles the ridge board, in that it receives the incoming shortened rafters, called jack rafters. If purlins are employed, they may cantilever from the nearest point of support to the hip. If the hip rafter does receive purlin load, it may be of large section, and exert thrust on the wall at its foot, displacement being prevented by a diagonal tie across the wall corner. Half trussed rafters may be used to carry purlins across the hipped end of the roof.

Valleys occur where pitched roofs intersect over the junctions between rectangular plan elements. Where roofs are tiled or slated, intersecting planes should be of identical pitch, to preserve the coursing of the roofing units. If the incoming jack rafters are carried on a complete main roof framing, they are fixed to a rafter board laid on the main rafters, but where no wall or beam is available to carry these main rafters, a *valley rafter* transfers the loads of the cut main rafters and jack rafters to the external wall (Figs. 9e, f).

The diagrams in Fig. 9 illustrate the constructional details described above.

*Wind effects*
Positive wind pressures on roof surfaces tend to strip off roof coverings composed of small units such as slates and tiles. They also exert "lift" on eaves and verge overhangs, due to rising wind spread off walls (Fig. 4). Negative pressure, commonly known as "suction," is developed on the lee side of pitched roofs, and may be sufficient to pull corrugated sheet roofing over its fixings, if the washer area under the nut of the fixing bolt is inadequate. Several notable cases have occurred of the stripping of whole roof structures.

Lightweight roof systems are very vulnerable to lifting by suction. In particular, monopitches with wide overhangs exposed to strong winds and gusting are highly susceptible and must be firmly tied down. The extent of this tying down will require careful assessment and calculation in large, light and exposed roofs. As an empirical example, on a short-span, low-pitch timber roof covered with metal or bituminous felt on decking, steel ties of about 25 by 3 mm section at 2 m centres would be turned over the wall-plate, carried at least 610 mm down the wall and inserted into a mortar bed joint. Preferably, the ties would be carried over or through rafters. The weight of wall above the final turn-in of the metal ties resists the lifting action.

Where long timber roofs are unrestrained by party or gable walls, wind may also cause racking or distortion of the framing. This can be resisted by fixing flat timber diagonal bracing, about 150 by 25 mm in section, to the undersides of the rafters, at intervals. Diagonal wind bracing at rafter level is often provided between trusses. Rigidity can also be ensured by the use of deep lattice purlin connections, or a pattern of diagonal bracing between trusses, connecting the top and bottom chords of adjacent trusses.

*Medium to large span trusses* (Figs. 10, 11, 12)
For a wide range of linear roof structures (for example, trusses, arches and rigid frames), the principle of collection of live and dead loads remains essentially the same. The covering, or its underlying deck, spans 2 or 3 m between purlins. These in turn span, from 3 m upward, on to the primary structural members, which collect a series of purlin point loads and transmit them to walls, columns, or ground. In steel trusses, where primary members are closely spaced, and the purlin spans short, small standard hot-formed sections, angles, and tubes, are adequate and economical for the whole system (Figs. 12a, b). Cold-rolled sections using steel sheet offer lightness and material economy for short spans, and may be formed into a variety of functional profiles. Cold-rolled Z section purlins are a good example, and are available in a catalogued range (Fig. 12c). Sag bars are used between supports to prevent lateral deflection down the roof plane.

If a standard hot-rolled section is too heavy or structurally uneconomic, a castellated (Fig. 19j) or trussed purlin can be used. In timber construction, short-span purlins are of solid sawn or planed softwood. For intermediate spans, plywood box or I sections are employed (Figs. 12d, e). Where long-span trusses or frames are at 10 m centres or more connectored, nail-plate, or glued trussed purlin or solid laminated members will be required (Fig. 8a). The effective purlin span can be reduced by diagonal strutting from the bottom chord of a deep truss (cf. the purlin strutted off crosswalls, Fig. 8b).

Triangulated trusses similar in form to the domestic version already described have long been in wide use for medium spans. They are amenable to straightforward structural design procedures and fabrication, in both timber and steel. Lattice trusses and girders are also constructed in prestressed precast concrete. These are pretensioned and precast as a complete truss, or are precast as a series of paired or matched sections assembled with mortared joints

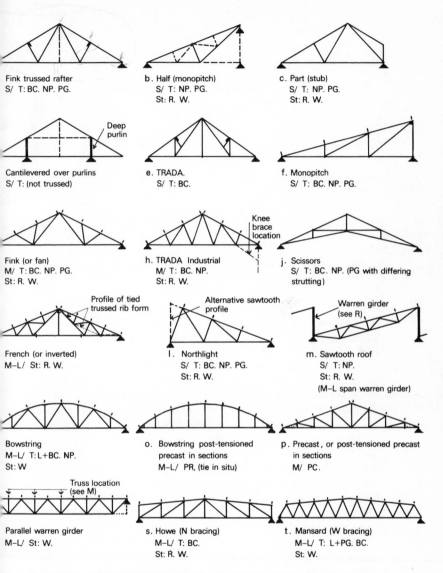

Fink trussed rafter
S/ T: BC. NP. PG.

b. Half (monopitch)
S/ T: NP. PG.
St: R. W.

c. Part (stub)
S/ T: NP. PG.
St: R. W.

Deep purlin

Cantilevered over purlins
S/ T: (not trussed)

e. TRADA.
S/ T: BC.

f. Monopitch
S/ T: BC. NP. PG.

Fink (or fan)
M/ T: BC. NP. PG.
St: R. W.

Knee brace location

h. TRADA Industrial
M/ T: BC. NP.
St: R. W.

j. Scissors
S/ T: BC. NP. (PG with differing strutting)

Profile of tied trussed rib form

French (or inverted)
M–L/ St: R. W.

Alternative sawtooth profile

l. Northlight
S/ T: BC. NP. PG.
St: R. W.

Warren girder (see R)

m. Sawtooth roof
S/ T: NP.
St: R. W.
(M–L span warren girder)

Bowstring
M–L/ T: L+BC. NP.
St: W

o. Bowstring post-tensioned precast in sections
M–L/ PR, (tie in situ)

p. Precast, or post-tensioned precast in sections
M/ PC.

Truss location (see M)

Parallel warren girder
M–L/ St: W.

s. Howe (N bracing)
M–L/ T: BC.
St: R. W.

t. Mansard (W bracing)
M–L/ T: L+PG. BC.
St: W.

Figure 10. TYPICAL SIMPLY SUPPORTED PLANE LATTICE TRUSS AND GIRDER FORMS

CODE:    Span Range, as shown:
         S/= Short, up to 10 m
         M/= Medium, from 10–20 m
         L/= Long, above 20 m

Typical Materials and Techniques, as shown:
         T. =timber
         BC.=bolted connectored
         NP.=nail plate
          L. =laminated
         PG.=ply gusseted
         St.=steel
          R. =rivetted
          W.=welded
         PR.=precast concrete

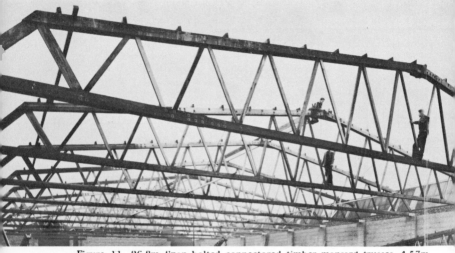

Figure 11. 26.8m Span bolted connectored timber mansard trusses, 4.57m centres

into a truss. The reinforcing steel strands are then tensioned to pre-compress the concrete units (Fig. 10p).

Economy in lattice trusses results from the elimination of redundant material from the basic solid beam form, resulting in a network of members in direct tension and compression. The pitch of a triangulated truss is related to that required for the covering or decking (where this is critical) and also to design economy. A double-pitched truss has maximum structural depth at or near its mid-span, where bending moments are greatest. The layout of struts and ties is usually related to purlin spacing. Typical truss forms are shown in Fig. 10. Note that certain medium-span trusses embody a pair of inverted symmetrical trusses, connected at the ridge, and by a central tie and hanger in an inverted T form (Fig. 10k). This makes for ease of fabrication of identical units, and for the simpler transportation of half trusses.

Support at truss feet may be provided by precast concrete padstones, transmitting a half truss load on to an area of loadbearing walling, or by a steel cap plate at the top of a column or stanchion. A knee-brace may be introduced at the truss-column junction forming a rigid wind brace (Fig. 10h). At the column base, a non-rigid joint prevents rotation in the foundations. The weight and area of the foundation pad must be adequate to prevent uplift, as in severe wind conditions the rigid truss and column frame tends to pivot on the lee side column foot. In the Howe truss (Fig. 10s) rigidity is achieved by fixing the vertical truss ends to the stanchions.

Although triangular lattice trusses may be used for spans up to 60 m, the air volume above eaves level which requires heating

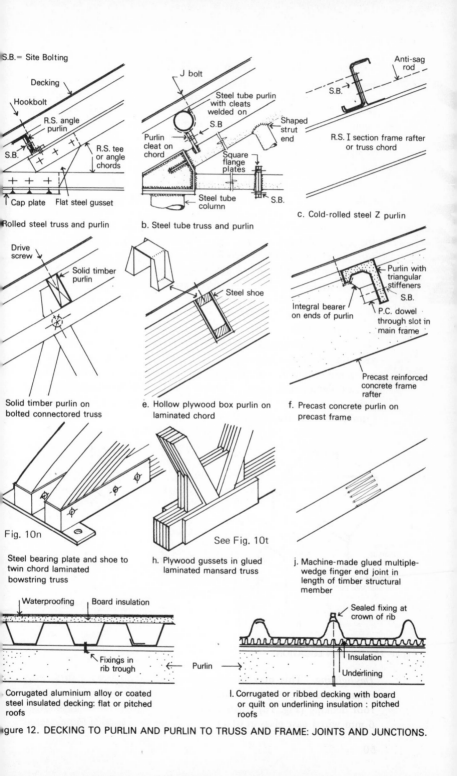

S.B. = Site Bolting

**a. Rolled steel truss and purlin**
Decking
Hookbolt
R.S. angle purlin
S.B.
R.S. tee or angle chords
Cap plate
Flat steel gusset

**b. Steel tube truss and purlin**
J bolt
Steel tube purlin with cleats welded on
S.B
Purlin cleat on chord
Shaped strut end
Square flange plates
S.B.
Steel tube column

**c. Cold-rolled steel Z purlin**
Anti-sag rod
S.B.
R.S. I section frame rafter or truss chord

**Solid timber purlin on bolted connectored truss**
Drive screw
Solid timber purlin

**e. Hollow plywood box purlin on laminated chord**
Steel shoe

**f. Precast concrete purlin on precast frame**
Purlin with triangular stiffeners
S.B.
Integral bearer on ends of purlin
P.C. dowel through slot in main frame
Precast reinforced concrete frame rafter

Fig. 10n

**Steel bearing plate and shoe to twin chord laminated bowstring truss**

See Fig. 10t

**h. Plywood gussets in glued laminated mansard truss**

**j. Machine-made glued multiple-wedge finger end joint in length of timber structural member**

Waterproofing    Board insulation
Fixings in rib trough
← Purlin →

**Corrugated aluminium alloy or coated steel insulated decking: flat or pitched roofs**

Sealed fixing at crown of rib
Insulation
Underlining

**l. Corrugated or ribbed decking with board or quilt on underlining insulation : pitched roofs**

Figure 12. DECKING TO PURLIN AND PURLIN TO TRUSS AND FRAME: JOINTS AND JUNCTIONS.

becomes very large, although clear height is restricted by the lower chord, and visually the roof space is occupied by a dense network of members. Maintenance painting of the truss surfaces is difficult, owing to the considerable rise, and the multiple surfaces of steel angle and tee are laborious and costly to paint. An insulated ceiling considerably reduces the heating problem and the visual disorder. Medium and long-span industrial roofs are usually fully exposed and admit daylight, so that a clean appearance is important. A portal frame is usually preferable to complex lattice trusses for very long spans.

*Column spacing* (see Fig. 13). If a *permanent* partition occurs under the valley between trusses, then columns under each truss foot, in the line of the partition, would be satisfactory. If, however, a large column grid dimension is necessary for maximum uninterrupted internal space, valley beams will permit the omission of alternate columns, and will carry every other truss at mid-span (Fig. 13a). Columns are retained under each truss at external walls, unless probable future extensions dictate otherwise. Where lightweight sheets are used as wall cladding, they are supported on horizontal rails (cf. purlins) spanning between the columns. If longer internal spans are required, with medium span triangular or diamond-shaped trusses, deep parallel lattice Warren girders can be introduced at the point of maximum depth. Such trusses form a double cantilevered "umbrella", the truss feet being linked by a light tie (Fig. 13b).

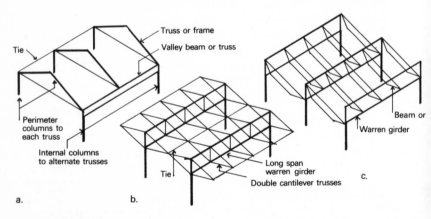

Figure 13. COLUMN LOCATION

*Daylighting* This may be introduced at the outer roof surface by clear or translucent plastic corrugated sheeting profiled to match the roof decking (Fig. 16). Also widely used is patent glazing, in which 6 mm wired cast glass in 610 mm wide panels is carried in protected or durable metal tee-shaped structural bars. If direct sun penetration is undesirable, an asymmetrical northlight or saw-tooth truss may be used, the steep or vertical plane being glazed (Figs. 10*l*, 17). As the span of these is usually about 8 m, a deep Warren girder is again introduced in the full depth of the truss, vertically or on the glazing plane (Fig. 10r). A castellated or light lattice beam may replace the northlight truss, connecting the top of one Warren girder with the bottom of the next (Fig. 10m). This system reduces the amount of visible structural network in the roof space. A further variant which eliminates roof space interruption has two parallel lattice girders inclined together to form a lattice plate, a three-dimensional form discussed later.

*The monitor roof* has two vertical or steeply-inclined glazed upstands in the monitor, a rectangular superstructure which is either constructed with a light framework on a main structural truss or beam, or as part of a cranked portal frame (Figs. 14h, 17).

*The Howe truss* (Fig. 10s) is designed for low pitches of about 5 to 10 deg, and for medium to long spans. The *Mansard* version of the Howe truss has steeply-pitched sides, often glazed (Figs. 10t 11). Both have low roof volume above the bottom chord.

*The Bowstring truss* (Figs. 1, 10n) has a continuously curved top chord and spans from 20 to 45 m, in timber and up to 60 m in steel, at truss centres of 4 to 7 m. In bolted connectored timber form, laminated timber top chords are used (Fig. 12g). In steel, tube sections are continuously curved or faceted to follow the roof profile. In timber the truss depth is about one-eighth of the span, in steel one-tenth to one-twentieth. With large radius curvatures flat decking may be used in a series of facets, but the decking, if exposed, must be capable of being used flat, as there is no pitch at the crown. If the building is occupied and heated the decking must be well insulated, to prevent condensation gathering on the central flat part of the soffit and dripping into the interior.

*Steel in lattice trusses*  Hot-rolled steel angles, tees, channels and I sections are very commonly used in bolted and riveted work, flat gusset plates being cut from steel sheet. The sub-assemblies are shop riveted, and site-bolted joints are made on the ground or in the final position. The face-to-face joints at the gussetted nodes are simple but rather crude in appearance. Circular steel tubes are structurally efficient, neat, and easy to paint. Steel tube trusses are normally shop-welded, with flange plates bolted on site, or, where repetition justifies provision of the necessary equipment and inspection facilities, are site welded. Tube junctions involve machining of tube ends to a curved profile to saddle on to the larger chord tubes, with

51

fillet welding, and the use of steel sheet welded up to provide the junction with wall or column. Tubular purlins meet at upstand cleats, fillet welded to the top chord of the truss. End-to-end butt joints, although very clean in appearance, are expensive, requiring precise splay machining to the tube ends, and a backing ring inserted into the tube ends, to contain the weld metal in the vee groove (Fig. 12b).

Rectangular steel tubes have increased in popularity for networks. The machining and fillet welding of adjacent surfaces is simpler than with circular or elliptical tubes.

*Timber in lattice trusses*   Long span trusses can be fabricated with multiple split-ring bolted connectored joints and nodes, single struts and ties entering twin top and bottom chords. As mentioned earlier, plywood gussets may be used in glued joints, interleaved with laminated members (Fig. 12h).

*Glued laminated timber construction*, also known as "glulam", is based upon the development of modern adhesives, and upon the stress grading of timber. Although simple casein glues can develop adequate strength under pressure, they are not waterproof, and are therefore unsuitable for external or damp locations. Urea-formaldehyde and resorcinol-resin adhesives are most commonly used.

Unlaminated solid timber construction has several limitations. Heavy sections of timber are difficult to obtain commercially in this country, the maximum section area and length available being limited. Drying of large timber is difficult and shrinkage and surface cracking occurs. It cannot be shaped or curved without wastage, and being homogeneous and a natural vegetable material, interior structural faults are not easily assessable.

In laminated construction, seasoned planks of 19 to 50 mm thickness are employed, usually in horizontal lamination (the Mansard truss is an exception). The thinner laminae are needed for small radius curves, and having more glue lines than the thicker planks are the most expensive. These planks are used for straight beams, girders, large-radius "bents" and arch ribs. The planks are stress graded, normally by visual inspection by skilled graders who allocate the sections to a working stress grade according to criteria laid down in *BS Code of Practice 112*. The highest grades are located in positions of highest stress in the member, the lower grades in less critical positions. The fabricator can inspect, grade, and allocate pieces according to their structural qualities or defects.

A system of mechanical stress grading has recently been introduced into this country. Timber sections are subjected to a point load over part of their length and the deflection measured. A computer interprets the information and allocates each piece to a stress grade category, whence it is diverted to an appropriate stack of timber. The machine is fed continuously, and takes account of the warp, bend and twist in each piece. It has a far greater output than any human grader, and is able to assess measured structural characteristics, rather than visual ones. Its use should lead to greater production of stress-graded timber, at much lower cost than is possible by visual means.

Laminated members of considerable length are produced by the use of staggered end joints of the laminae. In solid sections, machine made glued and tapered finger joints have increased available lengths and reduced waste (Fig. 12j). Although production of standard rectangular straight laminated beams (Fig. 19b) has now been started in this country, the labour and glues required in their construction means that for some time at least they will be more expensive than similar solid sections, although there may be structural and visual advantages in their use.

The laminated member is made by setting up a series of cramps in a jig to the required profile. Laminae are passed through an adhesive spreading machine, assembled in the jig, with any necessary curvature or taper, and cramped. The surfaces of the completed member are planed to complete the shaping. Considerable variation in the cost of the laminated member is possible, being affected by the quality of timbers and adhesives used, and upon the degree of surface finishing required. Purpose-made steel straps and shoes are used in conjunction with timber roof frames, for the connection of sections, and for fixing to structural supports (Figs. 12e, g).

The key letters to diagrams in Figure 10 identify those structures which are appropriate to the principal structural materials and techniques.

*Rigid frames*

A roof of lattice trusses or girders, carried at eaves level, has a large roof volume which is visually and physically obstructed, even comparatively simple arched-rib forms having visible ties and hangers. Unless knee-braced, the joint between truss and column is non-rigid, and the column does not contribute structurally by reducing bending stresses in the truss, which would otherwise yield economies of material.

A rigid, or portal frame (so called from its resemblance to a doorway), is characterized by a stiff eaves joint, which provides structural continuity between rafter and column. Bending stresses are taken in part by the column, the size of which is consequently greater than if it were only loaded axially, but the roof rafter may be a simple parallel or tapered member (Fig. 14).

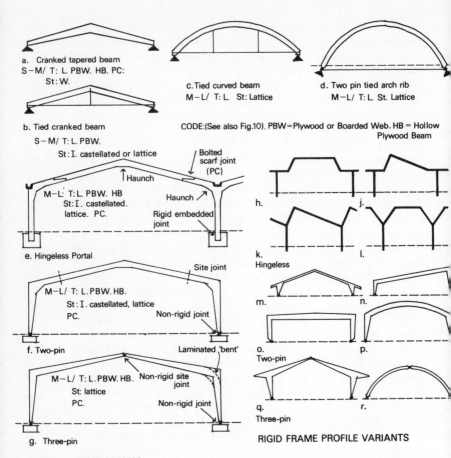

a. Cranked tapered beam
S−M/ T: L. PBW. HB. PC:
St: W.

b. Tied cranked beam
S−M/ T: L. PBW.

c. Tied curved beam
M−L/ T: L. St: Lattice

d. Two pin tied arch rib
M−L/ T: L. St. Lattice

CODE:(See also Fig.10). PBW=Plywood or Boarded Web. HB = Hollow
Plywood Beam

e. Hingeless Portal
St: I. castellated or lattice
Bolted scarf joint (PC)
Haunch
Haunch
Rigid embedded joint
M−L: T: L. PBW. HB
St: I. castellated. lattice. PC.

f. Two-pin
M−L/ T: L. PBW. HB.
St: I. castellated, lattice
PC.
Site joint
Non-rigid joint

g. Three-pin
M−L/ T: L. PBW. HB.
St: lattice
PC.
Laminated 'bent'
Non-rigid site joint
Non-rigid joint

h.    j.

k. Hingeless    l.

m.    n.

o. Two-pin    p.

q. Three-pin    r.

RIGID FRAME PROFILE VARIANTS

Figure 14. RIGID FRAMES

Structurally the most favourable arch form is the parabola, conforming to the curve of pressure, and thus, for normal live and dead loads, but not wind, the material in the arch is in a state of direct compression. Long-span enclosures for stadia are often of parabolic or continuously curved arched rib form (Fig. 14d and Fig. 15).

In the flat or pitched rigid frame, the greatest bending moments occur at the eaves junction, and the depth of the structural section is increased by means of a haunch, of splayed or curved shape, at this point. A rigid ridge joint will also be haunched.

*The hingeless portal* Structurally the most efficient and economical form is the fixed, hingeless portal frame, in which the columns are effectively fixed to, or embedded in the foundation blocks, and the ridge and any intermediate joints are rigid and

54

Figure 15. Laminated timber arch rib 94.48m span

structurally continuous. Design (by plastic theory) results in very slender rafter and column sections. Lifting and rotational wind effects may, however, be transmitted to the foundations, which must be capable of resisting them (Figs. 14e, 16).

*The two-pin frame* is produced by constructing "pin" or hinged joints, which may be actual rocker, axle or ball joints, or a designed weakening of the cross section, at the column feet. Bending moments are consequently not transferred to the foundations. The

Figure 16. Steel portal frame 41.14m span, 7.77m centres corrugated aluminium dec

Figure 17. Precast reinforced concrete rigid monitor frame

loss of negative bending moments produced in the hingeless portal by vertical cantilever from the rigid base joint results in a larger bending moment in the column. Thus, a larger section is required in the two-pin frame. A tension tie between eaves will reduce both the thrust of pitched rafters and the bending moment at the haunch (Figs. a–d, f, m–o).

In the hingeless and two-pin frame, bolted site joints are made in the rafters while the components are supported by staging or suspended by crane. These joints occur at the point of contraflexure, where the bending moments are at a minimum. The tensile stresses pass from the upper part of the rafter at the eaves, to the underside of the rafter at the ridge. A single-bay frame will have two identical outer sections and a central ridge section. In multi-bay frames, the interior columns form T or Y-sections with the incoming rafters.

*The three-pin frame* is hinged at the ridge or crown, and at the column feet. Structurally, this form is determinate and simpler to design, but less economical in material than those previously described. Often, the frame consists of two identical halves, with a single *in situ* bolted joint at the ridge. The provision of hinged feet facilitates lifting the frame to its final vertical position (Figs. 14g, q and r).

Typical frame shapes are shown in Fig. 14. The roof may be flat, monopitch, flat with cranked monitor, symmetrical or northlight double pitch, or have a curved roof profile. Spans are from 8 m up to 75 m, at from 4.5 m to 9 m centres. At the larger spans, lattice forms of steel, aluminium alloy and timber are used, sometimes rectangular or triangular lattice in cross section, to counter the buckling tendency of a thin linear frame. For small and medium spans, constant-section rafters and columns of I profile in hot-rolled steel are common. These are easily painted and maintained. Castellated or "Litzka" beams (Fig. 19j, k) are also used. The haunches require extra steel welded in, and their cost is affected by the amount of machining and shop welding of haunches required by the design. Rafter joints are often bolted and cleated butt joints, since welding in position at high level tends to be expensive. Splay cutting of the web and re-welding to form tapered steel members in three-pin frames is also costly in labour but mechanical welding can reduce the cost.

Precast reinforced-concrete frames of small and medium span are clean in appearance, require no maintenance if self-finished, and are fire resistant. To avoid moulding complexities and transportation difficulties which would arise if projecting purlin bearers were integrally cast, the frames have rectangular holes through which precast dowels are passed to receive the purlin ends. Precast purlins are inverted L or T-section, about 40 mm or 50 mm in thickness of

Figure 18. Precast reinforced concrete rigid northlight frame. Note columns under alternate frames

concrete, and with stiffening ribs at intervals in the length, bolted to the corbel dowels on the frame (Figs. 12f, 18.)

Haunches at the ridge and eaves of the frame are integral, and bolted scarf joints are used in the rafters. As the feet may be grouted into deep pockets in the foundation blocks, rigid portals are common. For large spans, dead weight may be reduced by coffering, and by tapering the section as the bending moment reduces. Steel hinge components are welded to the reinforcement before casting.

Several precasting companies offer a range of standard portal spans and pitches, for which they have set up the necessary mould facilities. For special designs requiring the provision of new moulds, rationalization of the number of different component patterns is important, as also is a degree of repetition. This minimizes the mould costs and spreads them over the number of precast parts produced for the project. Large-span portals create problems of transport and handling of heavy long units. Such frames are often cast horizontally on site with the column feet adjacent to their final position. The whole frame is then lifted through 90 deg into its ultimate location. The frames are designed for lifting, as well as for operational stresses.

Timber rigid frames are light in weight, easily transported, and durable, so long as timber species and adhesives are correctly selected for the conditions of exposure anticipated. Lattice and plywood box and web forms burn fairly readily, but heavy laminated members char slowly on the surface and remain stable under fire conditions.

In timber, several methods of constructing rigid frames are familiar, the best known being:

(a) *Plywood box*  Plywood outer webs are glued to a framework of solid or laminated timber flanges and struts. Haunches are straight splayed, segmented or curved, and the column and rafter usually tapers to steel hinge joints at ridge and floor. If left natural or clear finished, the selection of plywood for grain effect and joint pattern between sheets is important.

(b) *I section*  Flanges of solid timber are glued and sometimes nailed to a plywood web, so forming an I section. Stiffeners between the flanges stiffen the web against buckling. In the HB system (so called after the Swedish engineer Hilding Brosenius), vertically-laminated flanges are nailed to webs of layers of opposed diagonal boarding, giving a web of high shear resistance. Spans of 30 m are obtained economically.

(c) *Solid laminated*  Curved arch rib forms are possible up to spans of 60 m or more, using thick laminae of large radius to build up very large cross sections, without taper. The ribs will be spaced at about 10 m centres, with purlins of plywood box form, housed into steel shoes bearing on the arch rib (Fig. 15).

When used in pin-jointed portals, the boomerang-shaped "bent" forming a half frame has a tapering column and rafter, and a continuously curved eaves transition. If an eaves purlin is carried, the outer corner is built up with short laminations of diminishing length, which reduces the visual quality of the column-rafter junction. The taper is produced by planing the internal laminae, the outer ones being undiminished, so that feathered ends are not visible on the exterior of the member.

In all timber rigid frames, cost and appearance is related to type and qualities of species of softwood or hardwood, the type and grade of plywood, the strength and durability of adhesive, and to the quality and nature of any applied decorative and protective surface finishes.

# Chapter 4—Flat roofs

## Simple forms

The simplest forms of flat roof are those of short-span laid on loadbearing wall or beam and column frame structures. They consist of boarding on joists, or an *in situ* or precast reinforced concrete slab. The surface of these roofs must receive a waterproof covering, and where their inherent thermal insulation is inadequate, an insulating slab or board must be incorporated between the structural deck and covering.

*Solid lightweight precast reinforced concrete slabs* are produced by manufacturers of foamed concrete products (Fig. 5c). Low-density structural concrete has good thermal insulation, depending on the thickness employed, and the units may be used spanning directly on to precast concrete beams, portal frames or monitor frames, or on to purlins, so forming a deck of high fire resistance and durability. If there is likely to be interstitial condensation within the porous concrete, a continuous vapour barrier in the form of a sprayed pigmented plastic skin may be applied to the underside. Differential deflection between adjacent units is prevented by running a cement mortar grout into dovetail grooves formed by rebates cast in the edges of adjacent units. A splay on the lower arrises reduces the risk of damage during handling and masks slight differences of soffit level. If falls can be arranged by inclination of the structural deck itself, the waterproof covering is applied direct to the upper surface, without screeding.

*Flat short-span timber roofs* (Fig. 19a) utilise sawn softwood sections, usually not exceeding 225 by 50 mm, with an economic span limit of about 5.50 m, depending on joist spacing, and on decking, cover and superimposed loads. (See Tables 3 and 4 of Schedule 6 of the Building Regulations 1965).

Falls may be provided by tilting the joists or by nailing tapered sawn *firring pieces* to joists laid level. When fixed along the joist top, the minimum thickness of firrings may be 12 mm, but when laid across the joists the minimum depth of the firring piece must be adequate to span between them, and will be from 38 mm upwards. Excessive firring depth due to long one-way falls should be avoided as being wasteful of material and causing problems of fascia depth.

60

Note that firrings are formed by splay-sawing along a rectangular timber section. Thus a 100 by 50 mm section will yield two firrings tapering from 75 mm to 25 mm less the width of the saw cut.

If a timber-joisted flat roof is to have a ceiling, the joists should be selected for even depth, or re-sawn or planed to ensure that they are equal, to avoid ceiling undulations, since the top surface of the joist must be level. These surface irregularities are most apparent with a ceiling of single-coat plaster on plasterboard. Thin plaster does not allow for such unevenness to be concealed, as is possible to a certain degree with multi-coat plaster. Note also the need for ventilation of the sealed space between roof covering and ceiling in a timber flat roof.

If the joists of a timber flat roof are to be exposed in positions where appearance is important, they must be planed to identical section and free of bow or twist. Exposed surfaces between the joists will also require consideration.

*Deckings*† Sawn butt-edged softwood boarding may be used for low-cost work, but planed tongued-and-grooved boarding is better. Additional thermal insulation will normally be required on top of the boarding. Centres of joists will be 610 mm maximum for 25 mm nominal thickness boarding. A recent introduction from Canada has been multiple tongued-and-grooved softwood decking of 38 to 75 mm thickness, often Western Red Cedar. The decking spans upwards of 1.22 m, between heavy solid or laminated timber beams. To avoid waste at these thicknesses, end-to-end butt joints in the boards may be splined with short metal tongues, so that end jointing need not occur only over bearings. Using 75 mm softwood, thermal insulation may be adequate (Fig. 19b).

As an alternative to conventional strip boarding, plywood and chipboard roof decking is obtainable, with tongued-and-grooved edges. To obviate cutting to waste, large area sheets require a rigidly accurate joist layout, normally 406 mm or 610 mm for 1.22 m sheets, depending on spanning characteristics and thickness. Allowance must be made for dimension tolerances and expansion gaps between sheets. Both chipboard and strawboard are available pre-felted. With joints taped, a temporary waterproofing is provided before final completion of the covering. Pre-felting implies that any additional insulation required must be provided below the decking.

Structural insulating deckings, principally woodwool and compressed strawboard, are supported at 610 mm centres for a 50 mm thickness. Strawboard sheets run parallel to the joists, woodwool across them. The unsupported edges of the sheet deckings normally require *noggings*, small square timber sections, about 50 by 50 mm, fixed between joists to provide support.

† Most strip and sheet decking materials are equally suitable for flat and pitched roofs which have similar support structures (Figs. 3, 7f).

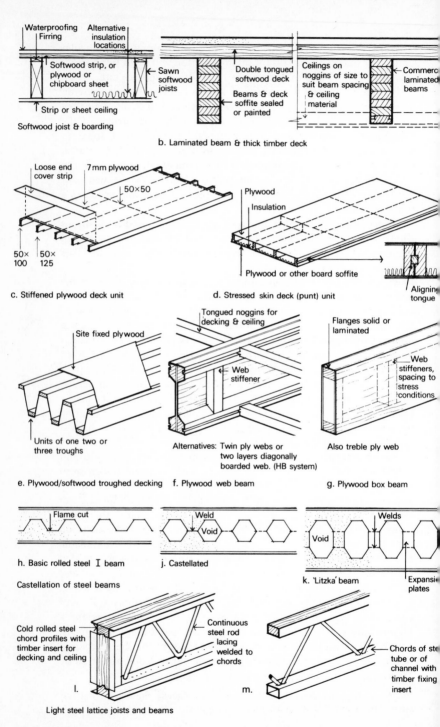

Figure 19. JOIST, BEAM AND DECKING ROOF COMPONENTS

A recently introduced "sandwich" deck has an upper layer of 4 mm or 6.5 mm birch plywood, bonded to a core of 19 or 25.4 mm rigid urethane foam. A coloured laminated polythene lower surface provides a vapour barrier. This lightweight structural sheet has high thermal insulation value, and spans 610 mm and 1.22 m in the two thicknesses.

The span capabilities of simple joists are calculated on the performance of the joist alone, ignoring any structural contribution made by decking or ceiling membranes, which are normally only nailed to the joist without consideration of their structural action. A sliding interface thus occurs when the joist deflects under load.

Substantial increases of span for unit depth of joist are possible if one or both membranes are properly bonded, nailed, screwed or machine stapled, to form *stiffened panels* (upper layer only) or *stressed-skin decks* (both layers). The panels use plywood as the upper surface, and the decks, plywood as the upper with plywood, asbestos insulation board, or sheet of adequate structural performance, as the lower surface, separated by planed softwood joists. Edge location and alignment, and mutual structural action, is provided by blocks engaging in grooves machined in the edge members of the unit (cf. tongued-and-grooved boarding). Additional insulation may be incorporated within the unit (Figs. 19c, d).

Deck units longer than the normal plywood sheet incorporate mechanically made scarf joints or multiple-finger joints at the ends of plywood sheets or timber members. Computer programmes have been prepared to help in the design of stressed-skin panels for a range of conditions. The technique is most suitable for large areas of roofing, for which factory jigs may be set up and continuous flow production maintained for the great number of units required.

Stiffened panels span from 2.50 m to 7.50 m, in depths from 50 to 300 mm. Stressed-skin panels span from 2.50 to 13.50 m for similar depths (Fig. 20). The plywood used is normally 7 to 12 mm thick. It will be seen that substantial spans may be achieved, and that the decking affords a clear plane soffit. Folded plate (Fig. 38) is also an appropriate hollow stressed-skin unit form.

A well known patent plywood deck using corrugated or folded configuration is "Trofdek" (Fig. 19e). Softwood upper and lower flanges are united by inclined plywood webs to form a "punt" of one or more corrugations in width. In the completed roof, an upper plywood layer receives the finish, and a ceiling may be applied if required.

*Other joist and beam systems*

The maximum span of sawn softwood joists for roofs is limited. In the deeper sections there may be significant drying shrinkage after

Figure 20. Plywood stressed-skin deck, 14-m span

completion of the building. The sections may not be identical in size unless planed, and bowing and twisting can occur unless strutting is used. Hitherto, large areas have been roofed by subdividing them with steel beams, carrying small timber joists. Within the last 15 years, a wide range of lightweight proprietary joists and beams in timber and steel has been developed, for spans beyond the economic and functional limit of standard sawn softwood sections. Spans of 15 m and beyond are possible, with depths up to 1.22 m. These joist-and-beam forms result from specialized structural design and manufacturing procedures, utilizing materials in advanced ways, in large-scale production of a range of sizes, upon which a price structure is based. Standard detailing for falls, openings, eaves and guttering is available, and special detail is often possible (Figs. 19f, g, l, m).

## Manufactured timber joists, beams and decking

### Plywood web beams
An I section is formed with webs of birch or fir plywood, stiffened against buckling under load by vertical softwood struts. Flanges are of profiled softwood sections, grooved to receive pre-cut tongued-end noggings between top and bottom flanges. The joists are bonded with waterproof glue. The noggings brace laterally and support both decking and ceiling. Maximum span is 14 m and the most economical spacing is at 1.22 m centres. The depth range is 254 mm to 1.22 m. Closer spacing of units may be adopted for high loadings, and twin units are used for trimming, the trimmed member being carried in steel shoes. Single or double falls are produced by firring. The units are pre-cambered at the larger spans to allow for

deflection under working load, and to achieve designed falls. Plywood web beams have the advantage over solid timber that there is little or no shrinkage in their depth (Fig. 19f).

In lightweight roofing systems, reduction of deadweight is achieved by the economical use of material in structurally non-critical areas, combined with substantial depth. The possibility of deflection under heavy snow load (even to the 0.003 of span allowed by British Standard Code of Practice no. 113 (The Structural Use of Timber)), makes it preferable to use movement-tolerant sheet, tile or strip ceiling finishes, as rigid plaster would crack, and the cracks would not be reparable by normal filling.

I-section beams, using webs of plywood or diagonally-opposed boarding layers, may be purpose-designed. The web stiffeners will be at varying centres, and located to suit the shear stresses occurring in the length of the beam.

An ingenious variation of the plywood web beam, illustrative of the effect of modern manufacturing techniques on structural components, has a self-stiffening plywood web corrugated in the length of the member. Plywood strip is fed from coils, spread with glue along the tapered edges, and inserted into grooves machine-cut to a sine curve in the softwood flanges. The beams are available in a range of depths between 300 mm and 600 mm, and in widths of 100 mm to 150 mm. Span range is 3 m to 15 m. The beams are supplied slightly over the required length for site cutting, and end stiffeners are site-fixed. As with rolled-steel I sections, these beams may be employed in pitched portal frames with a suitable haunch detail. Cambering may be incorporated in the length of the beam.

Parallel beams and girders with Warren ("continuous W") bracing, may be constructed with bolted connectored joints or plywood gussets. These techniques are appropriate to specially-designed long span roofs using heavy timber sections. Standard manufactured lattice timber parallel girders, with nail plate or steel-reinforced joints, are light units, using sections of only 75 mm width or so. Spacings and spans are similar to the systems described above. The lattice beam has the advantage that space for the passage of services is visible and usable without any cutting away of web material.

Solid laminated softwood parallel joists and beams are also available commercially, using thick laminae, and are grooved for lateral bracing by nogging (Fig. 19b). Plywood box-beam construction for flat roofs is usually purpose-made for beams or purlins (Fig. 19g). The similarity of construction between plywood box beams and plywood stressed-skin roof deck units will be apparent, and both are sometimes used in the same roof.

As most manufacturers of timber structural components are concerned principally with structural efficiency and economy of production, rather than with appearance, designers should acquaint themselves with the visual characteristics of the standard finish, the timber and plywood surface quality, any visible adhesive spread, and should enquire whether any special works treatment is available, when the units are to be exposed to view.

Lightness of weight commends the use of timber units for short to medium spans, when they may be manhandled into position. For longer spans, lightness is useful in making up transport loads, although length and shape are often the restricting factors. The employment of mobile cranes of suitable capacities makes it equally possible to handle timber, steel, and precast concrete.

## Steel beams and girders

For lightly loaded roofs, normal hot-rolled steel I-beams are relatively inefficient, as the depth required to limit deflection results in unnecessary and redundant self-weight. As the steel in the solid web is inefficiently stressed, there is an uneconomic use of material. *Castellated beams,* developed about twenty years ago, increase the depth of a standard section by up to 50% without use of extra material, but require an additional manufacturing process. The steel web of a conventional I-section is cut to a toothed profile by moving the beam horizontally under a vertical cutting flame. The projecting edges of each half are then welded together to form a deeper beam with hexagonal web apertures, through which service pipes and cables may pass. The castellated beam (Figs. 19h, j) may be used in any roof beam or frame system to which it is structurally appropriate.

A further increase of depth is obtained by the patent "Litzka" process, in which rectangular steel web expansion plates are welded between the beam projections, producing octagonal holes. Spans up to 36 m with lateral ties to the bottom flange are possible with a beam of 1.22 m depth. The top and bottom flange widths of castellated beams may differ, using beam halves from different I-sections (Fig. 19k).

### Steel lattice beams and girders
A range of short- to medium-span Warren-type latticed steel parallel beams are manufactured by several specialist companies. The basic configuration has a continuous W-form lacing of steel rod or tube between chords.

One system is based upon cold-rolled light-gauge steel sections for the chords. The top chord may have an inset preservative-treated softwood batten to receive direct nailing or screwing of decking. Alternatively it may be formed to provide a narrow groove along the top surface which grips special spiral-drive nails. Another system uses twin square rods for the lower chord, enclosing and being welded to the apices of the square-rod lacing. The upper chord is of steel channel, with flanges downstanding to receive decking hook bolts, or of steel angle carrying a triangular fixing fillet. A third manufacturer employs a flanged cold-rolled steel "top hat" section or a rectangular tube as the chord. Various end details provide for the beam to be underslung, with a bearing on the top chord, bottom chord fixing to supports, or vertical end fixing to stanchion or wall. Cambering or falls may be incorporated. Spans range up to 20 m, depths to 0.90 m. Spacings are from 610 mm to 3.05 m centres. All systems embody wide variations of beam-end and lateral and wind bracing fixings (Figs. 191, m).

Long-span parallel Warren girders for flat roofs are produced in riveted and gusseted rolled-steel forms, but very "clean" units are possible in welded circular and rectangular tube construction. Maximum transportable length is about 27 m. External chord dimensions in tube construction may be maintained by varying the wall thickness of tube in the length of the chord. The diagonal braces of a parallel Warren girder are of identical length and shape, so that this form lends itself to mechanical repetitive cutting and machining of the bracing. Circular tubes may be flattened at their ends to simplify the junction weld. Circular-to-rectangular, and rectangular-to-rectangular joints allow for simple fillet welding. Node joints coincide with purlin cleat locations thus avoiding chord bending by point loads from purlins. Site-welded junctions may be made with the aid of temporary bolted flanges. An an alternative, the permanent joint may be bolted. An advantage of tube construction is that tube interiors are sealed against corrosion by the structural welds.

The structural design of long-span lattice steel girders may suggest duplication of the top chord members, to resist compression, with the twin members being separated by short tube spacers. Lengthening of the spacers produces a laterally rigid three-dimensional space frame, of inverted triangular cross section if the bottom chord is a single member, trapezoidal or rectangular if it is duplicated.

## Concrete lattice girders

Precast concrete flat lattice girder systems have not yet been produced commercially for general use. Precast reinforced concrete purlins are made for flat roofs and are used with a short-span lightweight insulating deck such as channel-reinforced woodwool slab. One industrialized precast concrete structural frame system employs a post-tensioned concrete lattice girder, in which triangular precast units are assembled into a girder on the ground. This is then post-tensioned and lifted into position.

# Chapter 5—Three-dimensional structures

All the roof systems so far described have been linear forms, spanning in one direction, and having small breadth in relation to their span and depth. One-way spanning members require that the loads are carried at their ends or intermediately by beams, walls or regular columns. If they receive concentrated loads, special design provision is required, the load then being distributed along the length of the member. For long spans, the individual structural component will probably be very large, creating transportation problems.

A three-dimensional network roof structure may be classified as one-way spanning, or multi-way spanning. The lattice steel girder, of triangular latticed cross-section, has substantial structural width, but spans only in one direction on to supporting end stanchions. On the other hand, a space grid covers the plan area with a geometric network of members and spans in two or more directions on to a perimeter wall or perimeter columns.

*The space frame*
This may be regarded as a three-dimensional lattice truss or girder of great inherent lateral rigidity, by virtue of its braced triangular or trapezoidal cross section (Fig. 21b). Space frames may be underslung (that is, supported by their top chords) and spaced apart, so that a short-span deck may span between and across them. When upstanding, the clad surfaces of the space frame form pitched roof profiles. Space frames span only one way and require end support (Fig. 21d).

*The folded lattice plate roof*
When space frames are used for saw-tooth pitched roofs, the interior space is interrupted by horizontal diagonal bracing at the level of the bottom chord. If parallel Warren or N-braced girders are used, inclined to the pitches of the roof planes, the structure is confined to the roof surfaces (Fig. 21f). The structural depth available between ridge and valley is considerable, and spans of up to 90 m are possible provided that the structural depth is adequate. The folded lattice plate may be of northlight cross section, with glazing on the steeper plane, and with the ends hipped or gabled. Valley beams may

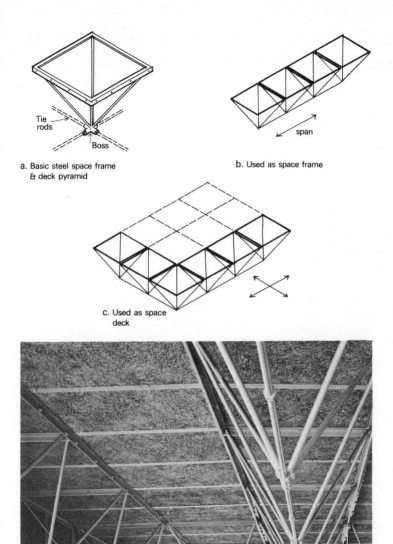

a. Basic steel space frame
   & deck pyramid

Tie rods

Boss

b. Used as space frame

span

c. Used as space
   deck

d. Steel space frame

Figure 21a-d. FOLDED LATTICE PLATE SPACE FRAME SPACE GRIDS

69

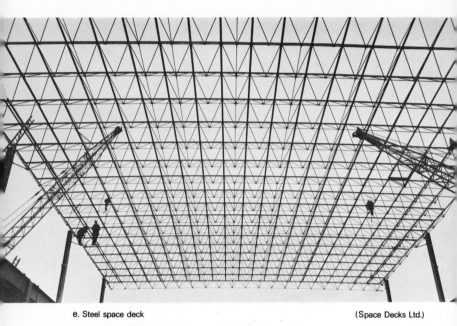

e. Steel space deck (Space Decks Ltd.)

f. Folded lattice plate

g. Single layer diagonal reinforced concrete grid

h. Double layer square lattice steel grid

Integral tie connector

Pyramid membrane of folded metal or moulded reinforced plastics. Sealed gutter flanges

Steel tube ties

j. Basic unit

Pyramid units on triangular, square or polygonal plan form

k. Pyramidal stressed skin space deck

Figure 21e-m.

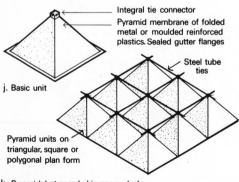

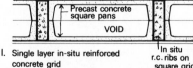

Precast concrete square pans

VOID

l. Single layer in-situ reinforced concrete grid

In situ r.c. ribs on square grid

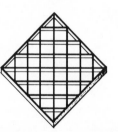

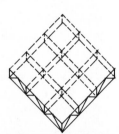

Tubes may enter at varying angles

Extruded aluminium cylindrical connector

Metal tube ends flattened & flared to slot into dovetail grooves in connector

Hole to receive bolt & washers to retain tubes in connector

m. Six way in situ joint in tubular metal space grid

70

be required for long spans, with perimeter beams containing the end plate against thrust. (Compare with surface-folded plates, Fig. 38.) A clean appearance is achieved with welded steel tube or rolled-steel fabrications.

### The single layer grid

Although not a true three-dimensional structure, the single-layer grid exhibits the characteristics of two-way spanning over medium distances, with moderate structural depth. A square, or near-square, plan is the most efficient, as the two spans are equal, or nearly so. The plan is divided into regular squares or diamonds, with sides about 1.22 m to 2 m. The single-layer grid is very appropriate for precast or *in situ* reinforced concrete, or steel framing over 12 to 20 m. The roof may be carried on continuous loadbearing walls, the load being transmitted downwards at frequent points by the structural ribs. Alternatively, a perimeter beam collects these loads and is supported by columns (Fig. 21g).

In *in situ* reinforced concrete grids, the rib form is created by casting concrete on to pressed steel or reinforced plastics pans, supported on flat shuttering. On striking, the pans are removed to leave a form of coffered roof. The slab thickness is about 75 to 100 mm. For special work, purpose-made timber and plywood boxes are prepared. The concrete may also be cast on to precast concrete or "ferro-cement" (reinforced fine concrete) pans which remain as the coffered soffit. Ribs may also be cast between concrete pans placed edge to edge. A flat soffit is thus produced, and the deep slab embodies weight-reducing voids between the pans (Fig. 21l).

In precast-concrete grids, straight members of a length equal to the side of one grid space are placed on bearers. Steel cables are passed through sleeves in these and are post-tensioned at the perimeter of the completed slab. Intersections between the rib sections are grouted up with a rapid-hardening mortar. The precast ribs may be provided with forked ends to increase the size of the "boss" available for grouting.

Steel single-layer grids have either plain or castellated I-section members, prepared for site welding at junctions and at the perimeter beams. Pressed steel rib forms have also been devised. In precast and steel grids of this type, the covering slab is not structurally integral, and may consist of any suitable form of short-span deck.

*Two- and three-way lattice grids* are formed by the intersection of parallel girders on square or triangular bays respectively (Fig. 21h). They lend themselves to the prefabrication of single-bay sections, rigidly fixed at the deep four-way and six-way intersections. One timber component manufacturer produces a form of two-way grid employing plywood web beams effective in two directions.

Structural continuity is provided at junctions by the use of cruciform steel straps fixed to the wood flanges.

## The double-layer grid

Double-layer flat space grids are created when three-dimensionally triangulated members connect a top layer network with a corresponding bottom layer, the diagonal members connecting staggered intersections. The two layers may have parallel or diagonally opposed members. A familiar commercial form has "squares over squares" (Figs. 21a–e) but "triangles over triangles", and "triangles with hexagons" are also employed. The structural characteristics of these grids are:

(a) A very rigid deck, created as a result of multi-way triangulation.

(b) Long spans, requiring only a moderate depth of structure (about 1.1 m to span up to about 40 m).

(c) Stresses very even with concentrated loads distributed in several directions. Although structurally indeterminate, computer-aided calculation has facilitated analysis and design.

(d) Large openings made possible without special strengthening of the grid.

(e) Supports widely spaced which, if necessary, may be irregular. The support may be at the top or the bottom layer.

(f) A rigidity of deck which lends itself to long cantilevers beyond the supports.

In the deep lattice grid, large void spaces are available for services. Although the grid structure may be underdrawn by a ceiling spanning between the bottom ties, the structure is often exposed for the visual quality of its delicate network. The top covering will be decking panels or sheets spanning between upper grid members. The basic "pyramid" may be of stressed-skin metal sheet or opaque, translucent or clear plastics, only one layer of linear ties being used (Figs. 21j, k). Alternatively, the top or outer surface only may be of stressed-sheet form, the diagonals and inner network layer being retained.

Computer calculation and the development of special node connectors are key features of space grid work. Dr. Mengeringhause has invented the German "Mero" connector, a steel sphere with multiple screw-threaded sockets receiving bolts fixed into the tapered ends of metal tube members. The American "Unistrut" system employs square steel tubes of identical length and cross-section, bolted into a standard pressed-steel connector, using a single bolt at each tube end. In the Canadian "Triodetic" connector method, the flattened crimped ends of structural tubes slide into the corresponding serrated slots of a cast metal boss. The crimping prevents withdrawal, and may be imparted to the tube over a wide range of angles (Fig. 21m).

The "Nodus" joint, recently developed in Britain, employs pins which pass through eyes at the ends of each incoming steel tube strut, enabling the struts to assume various angles at the node. This makes it a very versatile system.

The British "Space Decks" system (Fig. 21e) has as its principal component a standard pyramid, 1.20 m square on plan by 1.05 m deep. The square top layer is of rolled-steel angle, to which four steel rod diagonals are welded. These are united in a four-way threaded boss, which receives tie bars screwed in to form the bottom layer. The pyramids stack for transport and are assembled in position on site by bolting together all adjacent angles and bottom ties. On completion, a complete deck section is lifted to bearing level by mobile cranes. The standard units span up to 22 m when assembled as one-way spanning space frames, and 37.7 by 37.7 m as a deck.

## Domes and vaults

Structural grid and frame types employed for pitched and flat roofs are also applicable to domical and vaulted roofs. For a polygonal plan, three-pin portal frames (Fig. 14g) can be arranged to meet at the roof apex, thus creating a pyramidal roof. For part-spherical or parabolic domes, curved ribs are used as radial structural members (Fig. 22a). The identical ribs can be formed in laminated timber, curved hot-rolled steel I-sections, or precast reinforced concrete. Concentric purlins and rails support the outer cladding, and diagonal bracing across the panels formed by ribs and rails stiffens the framework. The concentration of incoming ribs at the crown may be taken by a special steel boss. Alternatively, a compression ring receives the thrust of the ribs. The central "eye" so formed is glazed, or may be roofed over by a system of smaller and fewer radial ribs (Fig. 22b).

The outer decking must be capable of covering double curvature, without excessive waste of material. Ribbed structures are erected by supporting the tops of radiating ribs on a scaffolding tower, which also affords access for making connections.

Many of the single and double-layer, and two- and three-way grids described for flat decks can be used for domical roofs, over circular or polygonal plans. The latter are created by cutting the edges of the sphere, and by modification of the geometry of the network members. Spherical domes lend themselves to the structural characteristics of glass-fibre reinforced polyester resin (GRP) panels. In this application they are used as single-layer flat or pyramidal panels, triangular or diamond shaped in plan, connected together at their edges. The panels are designed as stiff stressed-skin units. The high thermal movement of the plastic is accommodated by the spherical form. The colour, translucency or transparency of the

Figure 22a. Laminated timber ribbed dome. Note steel shoe and hinged joint

surface will depend on the selection of the materials (Frontispiece and Fig. 23).

Long-span domes have been constructed in single-layer grid form, using ferro-cement precast ribbed coffered pans, with *in situ* concrete jointing. Timber and pressed-steel linear components are used to produce "Lamella" vaults and domes (Fig. 24). Each lamella is joined at its ends and centre to adjacent ones to form a lozenge grid. For cylindrical vaults, the unit size remains constant. In domes the unit size diminishes towards the crown, but is constant for a single ring of units.

*Braced barrel vaults* (Fig. 25). A barrel vault is a single-curvature part cylinder, in which, geometrically speaking, a longitudinal straight line moves parallel to an axis from which the vault radius is struck. In lattice work there are two principal forms, apart from lamella and

Figure 22b. Compression ring and steel rod bracing to laminated timber ribbed dome

grids. In the first, diagonally-braced parallel girders are linked edge to edge to produce a prismatic folded plate, spanning the length of the vault between trussed or portal end frames. The end frames thus conform to the cross-section profile of the vault, stiffening the structure and carrying the plates. The vault may be tilted to provide northlight through a deep parallel truss that supports the upper edge of one cylinder and the lower edge of the adjacent one.

In the second configuration, curved ribs span between triangular lattice valley beams and straight longitudinal members form rectangular panels with diagonal rod bracing. The roof network for one bay may be assembled in position on the ground prior to lifting into its final location. Support columns can be widely spaced and will occur under the ends of valley beams or end frames. Welded steel tube and timber lattices are most common. When clad with a

Figure 23. Geodesic dome employing translucent reinforced plastics panels

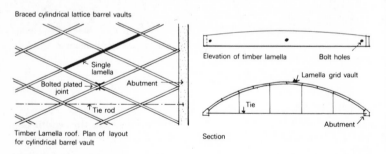

Braced cylindrical lattice barrel vaults

Single lamella

Bolted plated joint

Abutment

Tie rod

Timber Lamella roof. Plan of layout for cylindrical barrel vault

Elevation of timber lamella      Bolt holes

Lamella grid vault

Tie

Abutment

Section

Figure 24. Lamella roof

suitable sheet decking, the covering may be used to provide a stressed-skin effect. A steel network may also be enclosed in *in situ* concrete, but as a clad network, this form of "shell" is lighter than the reinforced concrete equivalent, providing an interior volume interrupted only by the end frames. To reduce lateral thrust at the valleys between the barrels, there should be a large rise (that is, a small radius) in relation to the width of the chord.

Figure 25. Steel lattice barrel vault, span 22.860m, rise 4.876m, embodying eight W-braced welded rectangular steel tube parallel trusses, welded edge to edge to form lattice vault. Carried by gable end trusses

## Tensile roofs

An essential of effective structural design is to attempt to use a material in the conditions of stress in which it is most efficient. Thus, concrete is ideally used in direct compression, and steel in linear members in tension. The suspension bridge illustrates the principle with regard to steel, its carriageway being suspended from vertical hangers attached to main cables hanging between towers in a natural catenary curve, and anchored to abutments that resist the pull of the cable. Roofs may be suspended from similar catenary cables, and long spans can be covered without visible internal structural members. It must be borne in mind, however, that maintenance costs of exposed towers and cables may be high.

In hanging roofs, steel cables or stranded ropes are suspended between perimeter compression rings of reinforced concrete or steel, the cables being in pure tension and with varying sag. Where the cables are radial, they converge upon a central tension ring. The network of cables may be clad with flexible plastic sheeting, sprayed plastics on a suitable background membrane, or with sprayed fine concrete on metal lathing clipped to the cables, prior to covering with flexible sheet waterproofing.

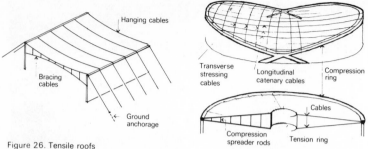

Figure 26. Tensile roofs

Suspended roofs can cover very long spans with very low self-weight per unit area. However, pure hanging membranes are subject to oscillation or "flutter" under varying suction. This inherent instability may be controlled by duplicating each cable vertically with stiff vertical spreaders (Fig. 26). These tension both cables, which radiate from a central hub and form the so-called "bicycle wheel" roof. This technique also provides an outward roof fall, overcoming a potential drainage problem if the sagging roof drains inwards to the centre. The hanging cables may also be pre-tensioned by stressing a second set of cables convexly at right angles over the first. A saddle shape is a typical form, with rainwater collecting at two low points. Within the requirements of anchorage, shape forming and stabilization, many roof variations are possible. Net-reinforced membranes draped over projecting masts, and double-pitched roofs stretched over a ridge beam are typical examples.

Metal tubes or lengths of lattice beam may also be used as suspended roofs, each length being hinged and linked end-to-end, in the form of a chain. In the case of the lattice system, a sag of one-tenth of the span is economical, and wind uplift is resisted by compression in the upper chords.

# SECTION THREE
# Surface structures for roofs

As used in roof construction, surface and membrane structures provide both spanning and covering functions. Loads on such structures are distributed in many directions, within the overall cover of the structural roof itself.

In this Section roofs based on *Shells and folded plates* (Chapter 5) and the comparatively recent types generically termed *Air-supported roofs* (Chapter 6) are considered.

# Chapter 6—Shells
# and folded plates

*A surface structure* is one in which the stresses due to dead and imposed loads are distributed within a thin enclosing membrane, and then transferred to its supports or to the ground. In the network or skeletal structure there is a sequence of load collection, from short-span deck to secondary members, thence to primary members, then to supports and finally to the ground. This sequence is seen in a typical truss and purlin roof. Load collection may also be through a system of short structural members, as in the space grid.

The essence of a surface roof is that its shape is a function of a structural requirement to place the enclosing loadbearing membrane as far as possible in direct stress, thus minimizing bending stresses and permitting small membrane thicknesses, in order to keep deadweight to a minimum.

In a masonry barrel vault supported upon continuous side walls, individual stones are in compression, the long arc of the vault being unable to pass through the shorter horizontal chord under gravity. The resultant thrust is countered by massive support walls, external buttressing or metal cross ties. Such a vault cannot span longitudinally, unless it is arched in its length, and sidewall openings consist of arcading, or piers and beams.

A comparable vault in *in situ* reinforced concrete is a homogeneous monolithic cylindrical shell (Fig. 2) the strength of which derives from its curvature, and which will span longitudinally as a trough-shaped beam, between columns at the four corners. The tendency of the thin shell to spread and sag at its lower edges is resisted either by thickening the concrete or forming a beam and introducing heavier reinforcement, which may be post-tensioned when spans exceed 30 m. The curvature of the shell is maintained and its lateral thrust accommodated by forming stiffening beams or diaphragms at intervals in the length. A cylindrical shell transmits direct load to its columns. This is illustrated by examples of cylindrical shells that have been cast and fully finished at ground level, the roof finally being jacked up into position. As the shell rises, precast interlocking column sections are introduced and post-tensioned. This technique saves much support and access scaffolding.

*The cylindrical shell* (Figs. 27a, b, 29, 30) is an example of a

81

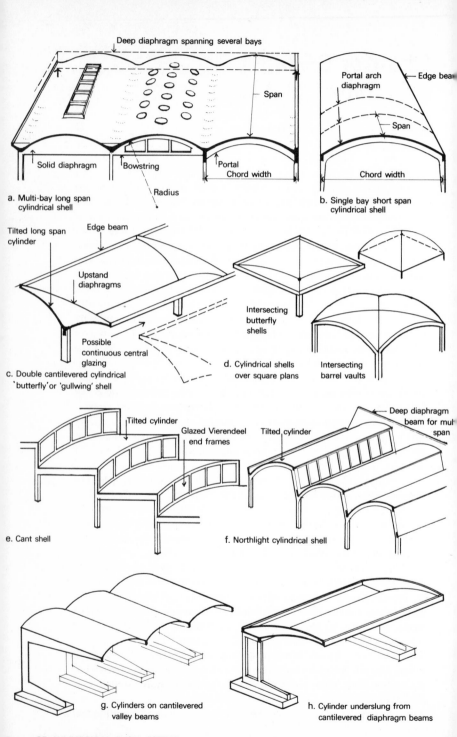

a. Multi-bay long span cylindrical shell

Deep diaphragm spanning several bays

Span

Solid diaphragm

Bowstring

Portal Chord width

Radius

b. Single bay short span cylindrical shell

Portal arch diaphragm

Edge beam

Span

Chord width

c. Double cantilevered cylindrical 'butterfly' or 'gullwing' shell

Tilted long span cylinder

Edge beam

Upstand diaphragms

Possible continuous central glazing

d. Cylindrical shells over square plans

Intersecting butterfly shells

Intersecting barrel vaults

e. Cant shell

Tilted cylinder

Glazed Vierendeel end frames

f. Northlight cylindrical shell

Tilted cylinder

Deep diaphragm beam for mul span

g. Cylinders on cantilevered valley beams

h. Cylinder underslung from cantilevered diaphragm beams

Figure 27. CYLINDRICAL SHELL FORMS

single curvature surface, in which a straight line passes over a curved one. It is the simplest form of shell, dating in its earliest manifestation from about 1925. There has been considerable development in post-war years. For economy, there should be repetition of the basic bay, forming a multi-bay cylindrical shell roof, in which the same shuttering is re-used several times. In countries where timber is abundant, purpose-made shuttering with plywood or boarded sheeting over runners carried on arched centering is common. In Britain, however, special shell-shuttering systems may be hired. Alternatively, a steel scaffolding contractor will erect a "birdcage" with steel sheets wired or clipped to horizontal tubular steel runners. The dismantling and re-erection of such a structure takes little time if efficiently done. Where the diaphragm beam is a portal rib, standing above the shell surface and leaving an uninterrupted soffit, mobile shuttering may be used. This is lowered, moved longitudinally and raised, bay by bay.

In roofs of this type, the edge or valley beam may be formed by thickening the edge of the shell in medium spans, or by a beam which will downstand at valleys, but it may be partially or fully upstand at the edges (Fig. 27a). The diaphragm may be solid, or perforated to form a bowstring, or a portal rib, with additional columns at end diaphragms to reduce the rib depth, if required. To reduce the area of the solid diaphragm internally, part of it may stand *above* the shell over the columns. This must, however, be perforated to accommodate the rainwater gutter in the valley, a detail which is often complicated to waterproof effectively.

The shell itself is thinnest at its crown at mid span, thickening towards the diaphragm and edge or valley beams. The soffit is normally a pure part-cylinder, the thickening being provided above. Thickness at the crown will be about 68 to 75 mm and may be determined not by compressive stress, but by minimum cover to reinforcement to prevent corrosion. The triangular spherical dome segments of New Hall, Cambridge (Fig. 28) are precast in ferro-cement, 12.20 m long, and only 22 mm thick, but zinc-coated steel-mesh reinforcement is employed.

Mild or high-yield stress steel welded square mesh reinforcement is used for the shell generally, with heavy diagonal rod reinforcement at the corners of the bay, and longitudinal rods at the lower edges, where tensile zones occur. The critical stresses and concentration of reinforcement restrict openings for lighting and ventilation to the central area, clear of the shell edges. Circular openings up to 1.20 m diameter have upstand stiffening rims that also serve as rooflight kerbs, and are staggered on plan (Fig. 29). Long rooflights parallel to the crown have rims, and cross ribs, to prevent inward buckling of the shell (Fig. 30). Cylindrical and domical shells are constructed with rectangular or circular glass lenses cast in, but severe sealing and

Figure 28. New Hall, Cambridge. Thin precast ferro-cement spherical shell
segments                                          (Crown copyright photograph.)

Figure 29. Single-bay cylindrical *in situ* reinforced concrete shell. Staggered
circular domelight openings with stiffening kerbs

Figure 30. Multibay cylindrical *in situ* reinforced concrete shell. Central
rooflight with stiffening ribs

84

waterproofing problems occur, and insulation is difficult to provide.

If a continuous central rooflight is required, double cantilevered cylindrical shells (Fig. 27c) are used. Being completely independent structurally, adjacent bays may be separated by any width required for the rooflight. It is the diaphragm or edge beam that is cantilevered, the shells rising from or being underslung beneath the cantilevered beams. The shell itself may be cantilevered beyond the end diaphragm.

*Long- and short-span cylindrical shells*   The long-span cylinder has a span to chord width ratio exceeding 3 to 2, with an optimum of 2 to 1. Span and chord range between 12.20 and 6.10 m to 30 and 15.25 m with normal reinforcement, beyond which valley and edge beams are normally prestressed.

Short-span cylindrical shells have a span to chord ratio of less than unity, and are used for single bay enclosures with a large span diaphragm. In fact they are usually rigid portals. The shell may be stiffened by intermediate ribs. A chord width of 61 m and a span of 15.25 m are the practical limits.

The radius of the cylinder should be carefully selected in order to provide a rise which is adequate for structural efficiency and economy without exeeding a 35 deg slope at the valley. Above this pitch, top shuttering may be required to contain the wet concrete. The shells are cast by laying fine (10 mm aggregate) rich concrete on the shuttering with careful spading and trowelling to profile. The standard of supervision and workmanship required in shell construction is high, and contractors experienced in such work, or specialist reinforced concrete constructors, should be employed.

*Northlight and cant shells*   Methods of admitting daylight through the shell cylinder have already been mentioned. A northlight roof is created by tilting each bay about its valley to admit daylight (Fig. 27f). For short spans the V of the valley may be adequately reinforced, the upper edge of the adjacent cylinder being propped by *in situ* or precast concrete posts at 2 to 3 m centres, but for long spans the whole northlight depth is used for a deep Vierendeel girder (that is, with rigid vertical post to chord connections and without diagonal bracing). A *cant shell roof* consists of short-span cylindrical shells inclined about the diaphragm to provide glazing (Fig. 27e).

Large square plan units can be roofed by means of two intersecting cylindrical shells. Three variations of intersecting cylinders over a square plan are shown in Figure 27d.

The single curvature shell is not restricted to a cylindrical cross section. The profile may be semi-elliptical or parabolic, and may rise from ground level, as in sports halls and other buildings for which the shape is very suitable.

Cylindrical shell roofs are conventionally supported by columns at each intersection of diaphragm and valley or edge, but use may be

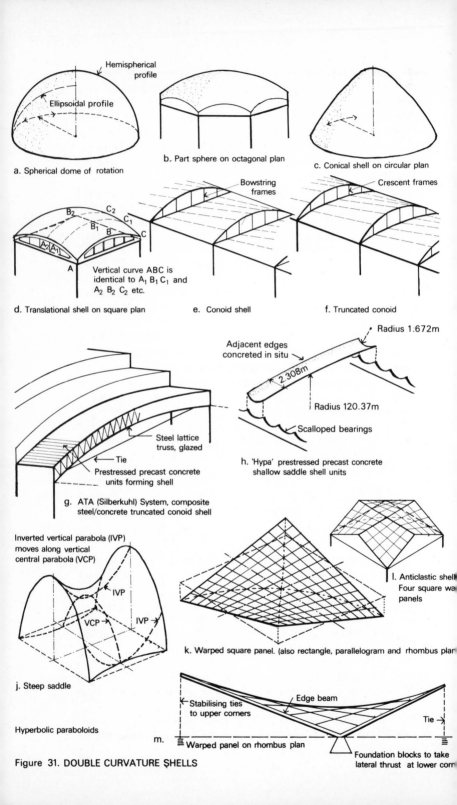

**a. Spherical dome of rotation**

Hemispherical profile

Ellipsoidal profile

**b. Part sphere on octagonal plan**

**c. Conical shell on circular plan**

**d. Translational shell on square plan**

$B_2$ $C_2$ $B_1$ $C_1$ $B$ $C$ $A_2$ $A_1$ $A$ Vertical curve ABC is identical to $A_1 B_1 C_1$ and $A_2 B_2 C_2$ etc.

**e. Conoid shell**

Bowstring frames

**f. Truncated conoid**

Crescent frames

**g. ATA (Silberkuhl) System, composite steel/concrete truncated conoid shell**

Steel lattice truss, glazed

Tie

Prestressed precast concrete units forming shell

**h. 'Hypa' prestressed precast concrete shallow saddle shell units**

Radius 1.672m

Adjacent edges concreted in situ

2.308m

Radius 120.37m

Scalloped bearings

**j. Steep saddle**

Inverted vertical parabola (IVP) moves along vertical central parabola (VCP)

IVP

VCP

IVP

Hyperbolic paraboloids

**k. Warped square panel. (also rectangle, parallelogram and rhombus plan**

**l. Anticlastic shell Four square wa panels**

**m. Warped panel on rhombus plan**

Stabilising ties to upper corners

Edge beam

Tie

Foundation blocks to take lateral thrust at lower corn

Figure 31. DOUBLE CURVATURE SHELLS

made of the full rise of the shell as a deep beam. This permits the omission of columns, or very large column grid dimensions.

*Double-curved shells* (Fig. 31)    The spherical shell is referred to as a double-curvature rotational shell, its shape being traced by radii from an axis, each radius lying on a curved generator that is itself half the profile of the shell (Fig. 31a). In a complete hemisphere, the load at the springing of the shell is vertical. In part spheres, there are thrusts at the perimeter which are taken by a circular or facetted ring beam, whose section is related to the spacing of supports, if the shell is carried on columns. The rotational shell cross section may also be conical, or a half ellipse. In the latter case, the ring beam can be produced by a smooth change of section to thicken the edge of the shell. When placed over triangular, square or polygonal plans, the rise of the cut edge of the shell dome may be treated as a portal or, alternatively, as a vertical or inclined glazed bowstring.

*The translational shell* is described by a convex curved line passing over and at right angles to a second convex line. Although the diagram (Fig. 31d) shows a translational shell over a square plan, and two curved lines of identical radius, this shell type is used over long rectangular plan bays, and in this form its convex crown may be compared with the straight crown of the single-curvature cylindrical shell.

*The conoid shell* is yet another form of northlight roof (Fig. 31e). To define it, a straight line moves over and at right angles to a straight line and a parallel curved line (normally the arc of a circle). When several conoid shells are adjacent, the segment of a circle between them is used as a bowstring diaphragm. If the conoid form is cut at right angles to its rise, a *truncated conoid* shell results and the bowstring shape is modified to a crescent (Figs. 31f, g, 32). An attractive feature of the conoid is that although of double curvature, it is cast on to shuttering of straight boarding which runs in the direction of the shell span.

*A saddle shell* is formed by the passing of a concave curve over two parallel convex curves of the same or different radii, hence the reference to the form of a riding saddle. The construction of shuttering for shapes of true double curvature is expensive, as unlike the truncated conoid described above, the shutter face and its bearers must be formed to the curvatures of the shell soffit. Shell domes have been cast on soil mounds used as moulds. These are consolidated and graded to the required profiles and removed after concrete casting and hardening. In the patent "Ctesiphon" system of corrugated shells, concrete is applied to hessian fabric suspended from bearers.

Where there is repetition of identical units of suitable size, precasting on site or at works is feasible, using doubly-curved moulds at ground level. One company produces a standard precast

Figure 32. Composite steel/precast concrete truncated conoid shell, 45.72m span, 7.62m centres

prestressed saddle roof unit, of 2.30 m width of "saddle" up to 20.72 m in length. These "Hypa" saddle units may be used in various combinations and positions on prepared scalloped bearings, edge to edge junctions being united by the use of *in situ* concrete (Figs. 31h, 33).

The saddle shell may be designed as a *hyperbolic paraboloid* (sometimes abbreviated to "hypar"), so called from its geometry, in which some sections of the membrane are hyperbolas and other sections parabolas. A simple demonstration of the form is given by raising a pair of the opposite corners of a square plane. The two lines joining the pairs of opposite corners form parabolas, one convex, the other concave. Lines joining opposite sides and parallel to the edges are straight lines (Fig. 31k). The doubly-curved surface shell so produced can be analysed structurally as a series of arch sections in one direction, and as suspension sections in the opposite direction.

Apart from this facility in structural calculation, the hyperbolic paraboloid has the constructional virtue that, for *in situ* concrete, the primary and secondary bearers and planking of the shuttering are straight members, suitably inclined to create the required structural profile. Similarly, such shells may be constructed in timber by using two or three layers of straight narrow-strip boarding laid in a

Figure 33. Prestressed precast reinforced-concrete shallow-saddle shell units

diagonally opposed pattern. In *in situ* concrete h.p. shells, reinforcement is straight and evenly disposed. Ribbed expanded-metal lath may be used as a shutter face, on to which fine concrete is sprayed. A thin shell of constant thickness may span a bay of 30 m square, the shell perimeter being stiffened with beams. Plan forms are the square, rectangle, parallelogram or rhombus, and bays can be combined together edge-to-edge in a wide variety of roof shapes, either simply supported or cantilevered (Figs. 31k, l, m, 34, 35, 36).

The lower corners of a panel exert a diagonal outward thrust as it tends to flatten. This is countered by inclined foundations, if the roof springs from ground level, or by a tension tie placed beneath the ground slab. When carried by columns, these will be in buttress form, or a visible tie will connect the opposite corners (Fig. 31m). In multibay forms thrusts may be arranged to balance between

Figure 34. *In situ* reinforced concrete hyperbolic paraboloid shell

Figure 35. *In situ* reinforced concrete hyperbolic paraboloid shell

Figure 36. Timber hyperbolic paraboloid roof, 12.20 by 12.20 m. Three layers softwood boarding, and laminated edge beams

Figure 37. Translucent reinforced plastics hyperbolic paraboloid modules 7.62 m span

panels. The tendency of a single h.p. shell to pivot about its lower corners is resisted by the provision of vertical ties from the corners to ground. Rainwater run-off concentrates at the lower corners and projecting chutes at these points, discharging over gratings at ground level, have often been used to dispose of the roof water.

Most shell roofs are easy to drain to their edges or ends, and may also have large-capacity valleys that act as gutters. Rainwater pipes with sealed joints can be cast into columns, which entails an increase in column cross-section. Alternatively, the down-pipe may be placed adjacent to the column, passing through reinforcement diversion in the corner of the shell. Where the shell valley has no fall provided by its shape, sufficient screeding should be provided to outlets. Such screeds should allow for deflection or creep in the shell.

Thin concrete shells respond rapidly to temperature variation. In barrels and domes, thermal expansion, restrained by valley beams, adjacent shells or perimeter beams, will take place in the shell curvature. In a cylinder, cumulative expansion will occur longitudinally to an extent dependent upon the temperature range experienced in the structural shell and upon its length. The maximum length between joints capable of absorbing the expansion should not exceed about 30 m and in addition, adequate thermal insulation should be provided above the shell. This is to limit the temperature range in the shell caused by external climatic conditions. Cork slab, woodwool slab, expanded-plastics boards or foamed or lightweight aggregate screeds are commonly used. Surface treatment of the soffit with low density materials, such as vermiculite plaster or asbestos-fibre spray, to reduce the concentration of reflected sound beneath curved shells should not be relied upon as primary thermal insulation. Downward heat loss from the shell will be retarded by the insulant, and retention of heat in the structural surface will aggravate the movement problem. Whenever possible the roof covering should be white or very light in colour to provide solar heat reflection. Multi-layer bituminous felt, with the top layer mineral-surfaced, is a common and economical waterproofing. Single-layer plastics or synthetic rubber coverings are possible alternatives, and offer the possibility of neat detailing on complex curves.

*Timber in shell construction* (Fig. 36). As mentioned earlier, laminated strip boarding, glued and nailed in layers in opposed directions, is used in hyperbolic paraboloid shells. The technique is also appropriate for other single- and double-curved forms, such as cylinders and conoids. Boundary beams are laminated in order to build up the section, and steel ties are employed. A cylindrical vault form is also made in timber, with plywood stressed skins on framing or other lightweight core in prefabricated sections, placed end to end and spanning between laminated valley beams.

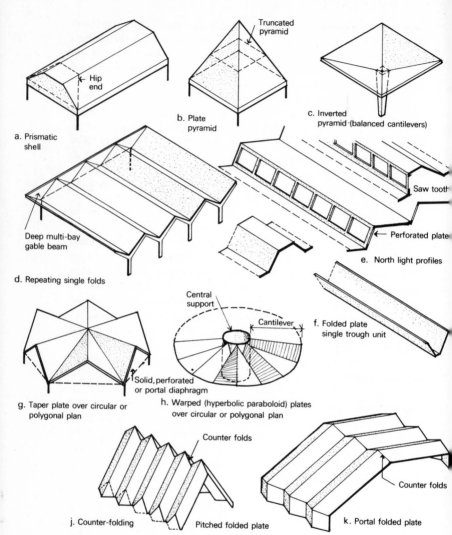

**a. Prismatic shell** — Hip end

**b. Plate pyramid** — Truncated pyramid

**c. Inverted pyramid·(balanced cantilevers)**

**d. Repeating single folds** — Deep multi-bay gable beam — Saw tooth — Perforated plate

**e. North light profiles**

**f. Folded plate single trough unit**

**g. Taper plate over circular or polygonal plan** — Solid, perforated or portal diaphragm

**h. Warped (hyperbolic paraboloid) plates over circular or polygonal plan** — Central support — Cantilever

**j. Counter-folding** — Counter folds — Pitched folded plate

**k. Portal folded plate** — Counter folds

Figure 38. FOLDED PLATE ROOFS

*Prismatic shells and folded plates* (Fig. 38). A high proportion of the total cost of curved concrete shells is taken up in the shuttering, which must be made up to curved profiles. Flat shuttering is simpler, and a lower cost results from dividing up a continuous curve (the cross section of a cylindrical shell, for example), into a series of equal straight facets or plates. The shell is then known as *prismatic*. The stresses in such a shell are partly transverse-bending across the

92

facets, between the stiffening folds of the junctions. The slab thickness of such a shell would be about 25 mm thicker than that required for an equivalent curved shell. A prismatic shell made up of identical folded slabs or plates can be cast *in situ*, or the plates precast and edge jointed with concrete *in situ*. Alternatively, the plates may be of hollow stressed-skin timber units.

By folding a slab of uniform thickness, a covering surface embodying increased structural depth is obtained, the gross area of the structural roof surface being larger than the net plan area covered. The plates transmit the imposed loads by spanning across the plate to the stiffening junctions, and the folded plate as a whole spans longitudinally between end bearings. Buckling of the thin, deep, inclined "beam" is prevented by the support of adjacent inclined plates.

Experiments with pleated cardboard as a model roof will show that when loaded, the unrestrained pleats tend to flatten, and the outermost ones to spread. In full-size structures, the folded plates are retained in position under load by diaphragm beams (Fig. 38). If these beams are the full depth of the folds, very long distances can be spanned. With portal frame stiffeners, columns occur under each valley (Fig. 38d, g). The end plates of a series are stiffened by an edge beam, or a valley beam is created by an upstand, or by a part plate.

The folded-plate principle has many variants. An asymmetrical system, with the shorter, more steeply-pitched plate constructed as a Vierendeel girder and glazed, forms a northlight roof having an uninterrupted internal surface. Z-shaped plates will create a monitor roof, and multiple plates the prismatic shell referred to above (Fig. 38e). V or U-shaped units (which may be precast if of a suitable span for lifting) may be used as a series of folded-beam units, edge-to-edge, or spaced out, with rooflighting or lightweight decking spanning between them (Fig. 38f).

The tendency of the folds to deform under load can be resisted by *counterfolding* (Fig. 38j). In the portal counterfolded-plate, the counterfolded sections are in reality walls, serrated on plan. The roof plates can also be counterfolded at the ridge to give a double pitch. Folded plates rising from a square or polygonal plan to an apex will form a pyramid, and this may be used inverted and supported by a central column as a balanced cantilever, draining to an outlet passing through the column. The maximum dimension possible at the base of the triangular plate should be limited to that which will permit a flat plate of an economical thickness, otherwise a stiffening rib pattern beneath the plate will be needed.

Triangular, or tapered plates, can be used over a polygonal or circular plan (Fig. 38g). "Warped" hyperbolic paraboloid plates are also used in this way, and can cantilever from a central drum or tubular column, the root of the cantilevered "umbrella" thus formed being the deepest part of the folds (Fig. 38h).

The width of a single plate in concrete should be such that no transverse stiffening ribs are required, as these would reduce the economy of the thin flat slab. The width of the unit (c.f. the chord width of a cylindrical shell) should be about one-third span, and should not exceed 9 m. The depth of the unit should be not less one-tenth span. The individual plate width is preferably limited to 3.66 m with a thickness of 100 mm. Plate slope should be limited to less than 35 deg, otherwise expensive top shuttering is required to retain the wet concrete and to prevent it from sliding down after casting.

The folded-plate form is an admirable vehicle for plywood stressed-skin applications up to about 18 m span. The edges of adjacent plates are splayed, and where they meet to form ridge, valley or fold, they are glued and bolted together.

To cast a complex counterfolded structure *in situ* would require expensive shuttering. In consequence, such structures should be designed, if possible, to employ a limited range of precast units, each having a single fold, and uniting into a whole counterfolded envelope, with *in situ* concrete at junctions.

The inherent stiffness of reinforced plastics, and their potentiality for repetitive moulding, make them very suitable for folded and counterfolded roofs and enclosures. Specialist moulders can now produce folded or hyperbolic-paraboloid "modules" in glass fibre-reinforced polyester resin materials. These are bolted together at their edges, each edge having a flange or U-channel shape which encloses or compresses a gasket or sealing compound. The material may be opaque, integrally coloured, or translucent for the admission of daylight (Fig. 37). Resistance to atmospheric corrosion is good. The relatively high thermal movement of the material is taken up by the folded geometry and thermal insulants are applied internally.

# Chapter 7—Air-supported roofs

In most structures, the dead and imposed loads are collected cumulatively within the loadbearing elements, and are then transmitted to the foundations, which distribute the stresses within the subsoil. In the air-supported (inflated) structure, on the other hand, the foundations are subject to tensile forces tending to withdraw them from the ground. This is because an air pressure greater than ambient atmospheric pressure is maintained within the external envelope to support it, with uniformly-distributed pressure on the enclosing membrane.

As the pressure is exerted normal to the surface in all directions, the natural shape of an air-supported structure is a sphere. In building terms, this is normally reduced to a three-quarter sphere or a hemisphere (Fig. 39a). In this shape the "inflatable" is often used to enclose radar equipment or exhibitions, for which a curved vertical section and circular plan is efficient. Two quarter-spheres may be connected to a half cylinder to make a long hall structure with parallel side and semi-circular ends (Fig. 39b). This shape is seen in market gardens as large greenhouses, polythene sheet being the material used.

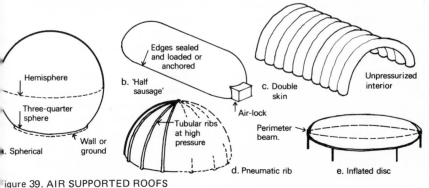

Figure 39. AIR SUPPORTED ROOFS

The structural membrane for nearly all air structures is usually a fully-flexible, impervious plastics fabric. Sharp edges between planes (as in a cube) are not practicable, since the edges crumple towards a regular curved shape. This may be observed if a rectangular

95

polythene bag is inflated. Thin steel or aluminium sheet are possible enclosing materials, but further development is required. The flexible envelope is in pure tension, stretched by the internal pressure. The curved shape is very stable under wind load, and adequate pressurization will resist "dimpling" and ultimate collapse.

The typical air-supported structure is inflated by electrically-powered fans, and outlet valves ensure regular air change and regulation of pressure. Minor punctures will not lead to collapse if the air pressure is adequate to compensate for leakage. The increase of air pressure internally does not cause physiological discomfort, being only of the order of 100 m of altitude. The structure is entered through an airlock, which may be in the form of revolving doors, which maintain an air seal.

Although clear polyethylene (polythene) sheet in normal heavy gauges is used for greenhouses, black polythene is more suitable for storage premises, being more durable than clear, as it degrades more slowly in sunlight. On the other hand solar heat absorption through a single-layer black membrane is high. Heavy plastic fabrics, such as nylon-reinforced vinyl with welded seams, are more durable, and are used for semi-permanent structures. A wide range of colour is available, providing varying amounts of light admission. Natural light is admitted by incorporating panels of transparent or translucent flexible plastics in the surface area.

The lower edge of the envelope is attached to strip foundations of sufficient mass to prevent uplift. In minor structures, the edge is simply folded into a perimeter trench, which is then backfilled with soil which loads and retains the fold.

The inflated shape may be modified or controlled by cables or nets, the ends of which are attached to anchorages of sufficient depth to resist withdrawal. Although the "inflatable" is often regarded as a complete wall and roof structure, some types of part-spherical inflated roof may rise from rigid walls of a circular plan building.

As the envelope is totally impermeable, the run-off water must be effectively collected around the perimeter. Snow is dispersed by curvature and by heat loss through the membrane. The membrane itself has very low thermal insulation value and gives poor reduction of air-borne sound. Double skin fabrics are available in which the two surfaces are spaced apart by multiple filaments bonded between them. With thermo-plastic films fire resistance of the bare surface is negligible.

Interior heating can be incorporated as part of the air-pump installation, or may be any other safe heating method. The interior sub-division of an air-supported superstructure will be arranged so that partitions do not contact the independent envelope. An attractive aspect of this type of structure is that it may be deflated and re-erected elsewhere with minimum labour requirement, if the packed unit is of suitable size and weight for carriage.

If a double-skin envelope is used, and the interspace inflated, a structure can be erected without a pressurised interior (Fig. 39c). Thus, it is possible to obtain a complete superstructure, or a roof alone as a pressurised convex disc, connected to perimeter walls or beam and columns (Fig 39e). Alternatively, the disc may be partially evacuated, as a "concave lens". Another air-supported roof not requiring internal pressure has ribs of tubular form inflated to high pressure. In such an application, they become stiff and support membranes stretched between them. This type will be familiar to campers, the small "pneumatic" tent with inflated corner ribs having been available for a number of years (Fig. 39d).

Although the inflated envelope is normally the "permanent" one, it may be used as a form work for a plastics or sprayed concrete shell, being deflated and removed when the shell has hardened.

The air-supported structure enjoys freedom from the need for structural support within the enclosure, requiring only air pressure to maintain its shape and stability. Immense inflated domes have been conceived, covering groups of buildings and even whole settlements. Within, building would be released from constraints of wind pressure and snow load, waterproofing, weathering and thermal insulation. Communities living in an equable climate could be created to exploit natural resources in, or to populate, areas previously considered uninhabitable because of adverse climatic conditions. The successful establishment of such roofed cities would depend upon the solution of major technological problems of environmental control, to maintain within such vast envelopes comfortable temperatures, optimum humidity and air kept free from accumulating pollutants.

With abundant power sources and mechanical installations of unprecedented capacity for environmental control, man may yet succeed in creating for the protection of his activities roofed enclosures of span, area and volume beyond his most advanced technical achievements of the present time.

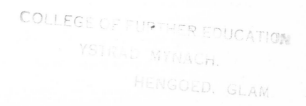

# Bibliography

*Mitchell's Elementary Building Construction.* Ed. R. Moxley. Batsford. London, 1959.
*Mitchell's Advanced Building Construction, The Structure.* Ed. J. S. Foster. Batsford. London, 1963.
Building Research Station. *Principles of Modern Building*, vol. 2. HMSO, 1961.
*Surface Structures in Building.* F. Angerer. Tiranti. London, 1961.
*Principles of Structural Design.* N. Lisborg. Batsford. London, 1961.
*Tensile Structures.* F. Otto. Volume I, 1967; Volume II, 1969. M.I.T. Press. Cambridge Mass.
*Frei Otto-Spanweiten.* C. Roland. Ulstein. Berlin, 1965.
*Principles of Pneumatic Architecture.* R. N. Dent, Architectural Press. London, 1971.
*Das Flache Dach.* W. Henn. Callweg. Munich, 1961.
*Building Elements.* R. L. Davies and D. J. Petty. Architectural Press. London, 1956.
*Structure in Building.* W. F. Cassie and J. H. Napper. Architectural Press. London, 1966.
Building Research Station. *Factory Building Studies 7, Structural Frameworks for Single-storey Factory Building.* HMSO, 1960.
*Steel Space Structures.* Z. S. Makowski. Michael Joseph. London, 1966.

*British Standard Codes of Practice:*
No. 3 Chapter V, Part 1. *Dead and Imposed Loads.*
No. 112. *The Structural Use of Timber.*
No. 114. *The Structural Use of Reinforced Concrete in Buildings.*
No. 115. *Structural Use of Prestressed Concrete in Buildings.*
No. 116. *Structural Use of Precast Concrete.*
British Standard 449. *Use of Structural Steel in Building.*

*Ministry of Public Building and Works*
Advisory Leaflets 55 and 56. *Timber Sizes for Small Buildings.* HMSO, 1962.

*Timber Research and Development Association*
(TRADA). *Timber Frame Housing Design Guide.* High Wycombe, 1966.
TRADA. *Standard Designs for Timber Trusses.*

*Thermal Insulation of Buildings.* G. D. Nash, J. Comrie and H. Broughton. HMSO, 1955.

*Factory Building Studies 11. Thermal Insulation of Factory Buildings.* G. D. Nash. HMSO, 1961.

*Building Research Station Digests*
(Second Series) HMSO:
No. 8. *Built-up Felt Roofs.*
No. 28. *Factory Building Studies.*
No. 51. *Developments in Roofing.*
Nos. 99, 100 and 105. *Wind Loading on Buildings.*
No. 107. *Roof Drainage.*
No. 110. *Condensation.*
No. 128. *Insulation against External Noise.*


*Truscon Reviews:*
No. 2. *Shell Roofs.*
No. 9. *Northlight Shell Roofs.*
No. 18. *Further Forms of Shell Roof.*
No. 29. *Folded Plates.*
No. 34. *Folded Plates 2.*


*Architects Journal Element Design Guides:*
Building Structure: General. 14.12.1966.
Building Structure: Metal. 15.2.1967.
Building Structure: Timber. 19.4.1967.
Building Structure: Concrete. 24.5.1967, 13.9.1967.
Building Enclosure: General. 30.8.1967.
Building Enclosure: Roofs. 25.10.1967, 15.11.1967, 29.11.1967.
*Architects Journal:* Flat Roof Failures. 30.6.71, 7.7.71.


*Regulations:*
*The Building Regulations 1965.*
*The Building Standards (Scotland) Regulations 1970.*
*The London Building Act: Constructional Byelaws 1967.*
*The Thermal Insulation (Industrial Buildings) Act 1957.*

# Index